William Earle Dodge Scott

Bird Studies

An account of the land birds of eastern North America

William Earle Dodge Scott

Bird Studies
An account of the land birds of eastern North America

ISBN/EAN: 9783337321956

Printed in Europe, USA, Canada, Australia, Japan

Cover: Foto ©berggeist007 / pixelio.de

More available books at **www.hansebooks.com**

BIRD STUDIES

AN ACCOUNT OF THE LAND BIRDS
OF EASTERN NORTH AMERICA

BY

WILLIAM E. D. SCOTT

WITH ILLUSTRATIONS FROM ORIGINAL
PHOTOGRAPHS

NEW YORK AND LONDON
G. P. PUTNAM'S SONS
The Knickerbocker Press
1898

To

The Memory of my Grandfather

JOSEPH WARREN SCOTT

OF

New Brunswick, New Jersey

Whose Ardent Love of Nature, Combined with his Tender
and Constant Care of its Unprotected Feathered
Tribes, Constituted one of the Strongest
Characteristics of his Heart and Mind

CONTENTS

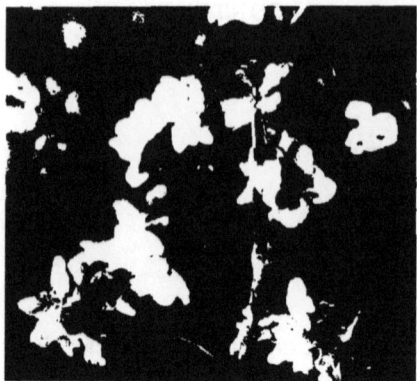

WHERE WARBLERS THRONG IN SPRINGTIME.

ILLUSTRATIONS.

Illustrations.

Illustrations.

Illustrations.

Illustrations.

A MARSHY GLADE ALONG A STREAM.

BIRD STUDIES.

INTRODUCTORY.

THE object of the treatise herewith presented is to place before students and others who wish to acquire knowledge on the subject, a means to that end. It is an invitation to a more intimate acquaintance with the Land Birds of Eastern North America. That is all.

By Eastern North America, as applied to the present subject, is meant that portion of the continent east of the Mississippi River, Lake Winnipeg, and the western borders of Hudson's Bay, together with Greenland and the islands which naturally group themselves with the mainland of the region.

The Land Birds include all birds found in this area, from the Song Birds, beginning with the Bluebirds, to the Gallinaceous Birds, ending with the Common Quail and its allied forms.

It is the intention of the author and the publishers, if this volume secures for itself an encouraging reception, to follow it with a second, dealing in a similar way with the Water Birds of the same region.

To point out to a would-be traveller a pleasant itinerary often induces him to take one route in preference to another. For this reason, systematic classification has not been adhered to in the following studies, but they have been made to group themselves about certain localities that are familiar to all. It is believed that a knowledge of the birds nearest to us is the best point of departure, and is less liable to lead to mental confusion, than if all the members of a given systematic group, as for instance all the thrushes, or all the sparrows of the entire region, were to be introduced or placed before the student in a body.

While it is not an easy task to select, from such a nomadic group as birds, the especial ones belonging to the environs of the house, yet certain kinds have come to associate themselves, on the whole, more with this than with any other locality. Others are in the same way characteristic of the woodland, the field and meadow, brook, marsh and swamp.

Our first problem will deal with those associated with the country house and its immediate vicinity. Thence we shall go by the highway to the woods and fields near by; and finally, we shall pass through the meadows, to the brooks and ponds with their marshy and swampy surroundings. There will be in each region thrushes, sparrows, warblers, wrens, and vireos, that are characteristic of that part of the journey. After each bird has become so familiar as to be known under all circumstances, an almost endless chain of research is opened, for the best known of our birds, even the Robin and humble Chipping Sparrow, will suggest many questions that have not been fully answered. What is their food, their period of incubation, the rapidity of their development,— in short, their life economy?

By word and picture the birds are portrayed. Two factors that are not very tangible, one song and the other color, will have to be acquired by observation and experience. It is easy to say that a bird is red, and that it has a charming song, but the exact tone and quality of the red has defied the brush of the painter, and no instrument made by man has yet rendered the music of the song. Form, however, is obvious, and the pattern of coloration can be shown. The shape of birds' bills and feet, wings and tail, together with their most salient markings, can be pictured with accuracy and described with certainty. This also is true in regard to birds' nests, the material of which they are constructed, and their contents, whether eggs or young.

Under the author's personal direction, Mr. A. R. Dugmore, with artistic perception and skill, has made the original photographs, which are carefully reproduced for this work. Some are taken from live birds, others from dead ones, some are from stuffed birds, others from prepared skins. All are faithful and accurate pictures, just what the camera presents, with its keen interpretation.

All of the live birds photographed for this book, and the greater number of the stuffed and mounted specimens, belong to the private collection of the author.

The birds' nests have been photographed in the places where they were found, and were in no way disturbed. So far as known, the parent birds of these nests had in no case abandoned them. Young birds were frequently photographed, some days after the nest and eggs had been utilized in a similar way.

In preparing the text, the author has made use of original notes based on field work and compiled during the past thirty years. The standard works on North American Birds have been carefully consulted to secure accuracy. It is believed that all the kinds of birds known to occur in the area dealt with, down to November first, 1897, are included in this treatise.

For the courtesies that have been extended to the author by Princeton University, in allowing him to use and photograph material in the museums of that institution, he desires to express his obligation.

WOOD THRUSH SETTING. PHOTOGRAPHED FROM LIFE.

ABOUT THE HOUSE.

BLUEBIRDS BUILDING.

ABOUT THE HOUSE.

IN the adult male Bluebird the entire upper parts including the wings and
tail are blue. Beneath, the front and sides are reddish brown. The
belly and feathers below the tail are white. The adult female bird
is much paler in color than the male, and the blue parts have a decided
grayish tinge. In both sexes the blue feathers are tipped
with rusty brown in autumn, and some individuals, presumably younger birds, retain more or less of these markings
in the spring. The adult birds are about seven inches in length.

Bluebird.
Sialia sialis (Linn.).

The young birds, on leaving the nest, have elongated whitish spots on
the blue of the back, which is of very pale shade. The feathers beneath are
grayish white and edged with black, so that the breast looks somewhat like
that of a Thrush.

In the breeding season Bluebirds are found in the United States, west
to the eastern slope of the Rocky Mountains, north to Manitoba and Nova
Scotia, south to the Gulf States. In winter they range from Southern New
York to the Gulf States.

It seems almost superfluous to introduce this bird to the student. Its
early arrival at the close of the winter, and the cheery note at a time when
bird life seems so limited in our Northern States, makes it at once conspicuous and welcome. Often before the snow has fairly melted its blithesome
voice is heard. Every old orchard, whose hollow trees afford nesting places,
soon has its resident pair. A deserted Woodpecker's hole, or the more
artificial "bird boxes," are not disdained. In such places a nest of fine
grasses is built and from four to six pale blue eggs are laid. There are
numerous records of sets of eggs pure white in color. They are more than
four fifths of an inch long, and exceed three fifths of an inch in their other
diameter. The time of breeding varies with the latitude from the middle of
March to May.

In winter the birds are gregarious. In flocks of from ten to fifty individuals they live a nomadic life, their wanderings being largely regulated by
the food supply and weather conditions.

The song is indicative of the season : its plaintive call passing overhead
in the autumn seems a harbinger of gray days and falling leaves ; and its gay

IMMATURE BLUEBIRD.

carol from the top of a fence post or from some leafless bough on the first promise of spring is surely the token of life and joy.

The head of an adult male Robin is black, and the eyelids and a spot a little in front of and above the eye are pure white. The bill is bright yellow, the extreme end being more or less blackish. The upper **American Robin.** part of the body is olive slate in coloring, and the tail Merula migratoria (Linn.). feathers are black, the outer feathers on each side being marked with white at their tips. The throat is white, streaked with blackish. The feathers below the tail are white at the tips, showing more or less of the slate color of their bases. The belly is also white. The rest of the lower parts and the sides are bright chestnut brown. The adult female is similar in color, but much paler or lighter, and with more slaty or gray feathers mixed with the black of the head, particularly on the back part. The birds are about ten inches in length.

The young on leaving the nest have the slate color of the back streaked with brownish white, and beneath are rusty white, spotted with black or slate color. In the autumn all the individuals have the chestnut color on the lower parts more or less obscured, each feather having an edging of white.

This bird is found in Eastern North America, west to the Rocky Mountains, and in Eastern Mexico and Alaska. It breeds from Virginia and Kansas north to the Arctic coast. It winters from Southern Canada, irregularly southward.

Like the Bluebirds, Robins are familiar and intimate with man. They nest by preference close to the house. Often the porch that has a suitable projection, or the nearest tree with a horizontal branch, affords a location for the nest. This is built, with an outer wall of coarse grasses and rootlets, woven into a mud wall carefully plastered, and lined with fine grasses. From three to five greenish blue eggs are laid. These are generally unspotted, but sometimes have brownish markings. They are about an inch and an eighth long, and nearly four fifths of an inch broad. The birds begin breeding from the last of March until early May, according to locality. Two broods are generally raised and sometimes three.

Robins are eminently gregarious, except in the breeding season. As soon as the first brood leaves the nest accompanied by the old males they nightly repair to some selected roosting place where hundreds or more congregate. These roosts are used till the falling leaves send most of the birds

ROBIN. OLD MALE IN AUTUMN.

ROBIN'S NEST.

II

to their winter homes.   Here, too, they are gregarious, and in Florida in
winter they wander through the woods in great flocks, reminding one of
Blackbirds.   Their food is largely berries of various kinds, such as those of
the mistletoe and holly; but in the north, though fruit is not disdained, insects and particularly worms are what the birds feed on.   Bird songs have
each their character, and to me the Robin's means good cheer and hope.

Wood Thrushes are brown above, brighter and with a cinnamon tinge
on head and shoulders, gradually becoming a clear olive brown on the
tail.   Beneath they are white, marked with clear round
Wood Thrush.   black spots, except on the feathers below the tail and
Turdus mustelinus Gmel.   on the belly.   The throat is white, defined by a clear
black line on either side running from the bill to the markings on breast.
The birds are about eight and a quarter inches in length.   The sexes are
alike.   The young, on leaving the nest, are much like adults, though the
feathers are more filamentous.   The brown feathers of the upper parts, head
and back, are streaked with dull buffish white.   The spotting below is more
transverse, and the white is suffused with buff on the sides and breast.
This Thrush is found in the Eastern United States, west to the Plains,
and as far north as Michigan, Ontario, and Massachusetts.   They winter in
Central America.   They breed from Virginia, Kentucky, and Kansas northward.   The nest is usually placed in a sapling ten feet from the ground.
The outside of the nest is of large dead leaves, roots, and twigs.   These are
woven into an inner wall of mud, and the lining is of fine rootlets.   The
eggs are two to five in number, are pale greenish blue in color, and unspotted.   They are a little more than an inch long and less than three quarters
of an inch broad.
This is the most conspicuous spotted Thrush, the only one that sings
commonly in the vicinity of New York and south of that point, except in
high altitudes.   About shady lawns and often close to houses, in the woods,
especially those that are damp and dark, he finds his summer home.   The
clear, bell-like notes that make his song so distinct in the summer orchestra
of bird life, particularly in the morning and evening, may be heard from the
piazza of almost any country house and in the woods near by.
The Wood Thrush comes to us late in April, or at the beginning of
May, and leaves in September and early October.   Individuals in the fall in
new unworn plumage are beautiful birds.   All the colors are intensified, and

WOOD THRUSH.

the texture of the feathers is finer even than in the spring. The moult is hardly completed when they leave us. At the time of year in question, the feathers on the back of the head, in old and young birds alike, have filamentous hair-like extensions to the quills that are very noticeable. In common with other thrushes, just before leaving these birds assemble in very considerable companies, feeding on gum-berries and other wild fruits.

WOOD THRUSH. ABOUT THREE WEEKS OLD.

The cold winter days bring to the house yard and orchard a gallant band of birds, animated, gay. inquisitive, and industrious, undaunted by the severest weather. Sometimes only a single pair is seen, but generally from four to ten Black-capped Chickadees form the party. In the autumn, after breeding and moulting are over, these little bands of Chickadees assemble and begin their winter cam-

Chickadee.
Parus atricapillus Linn.

15   WOOD THRUSH'S NEST.

paign against the dormant insect life. They are often accompanied by Golden-crowned Kinglets, and a pair of White or Red-bellied Nuthatches frequently join the gay company. These titmice are found north of the Potomac and Ohio valleys, and they breed from Southern Illinois and Pennsylvania northward, and on the summits of the Alleghanies, as far south as South Carolina.

The feathers of the Chickadee are of a peculiar downy fineness and are almost filamentous in character. The top of the head is black, which color extends down on the back of the neck to the back. The throat is also black. The sides of the head and neck are white. The breast and belly are white, and the sides and flanks are pale ash, washed with light buff. The back is ash color, the wings and tail feathers are a much darker shade of the same color, and are edged with white. The outer webs of the *larger feathers of the shoulders are also conspicuously edged with white.* The sexes do not differ in appearance, and the young, on leaving the nest, resemble the old birds. The birds are about five inches in length.

The nest is made of grasses, mosses, feathers, and similar soft material, and is placed in a hole, often excavated by the birds, in a post, stump, or tree trunk, usually not more than ten or fifteen feet from the ground. Five to eight eggs are laid. They are white, spotted with reddish brown, particularly at the larger end, and are about three fifths of an inch long, and not quite half an inch in their other diameter.

The call of the Chickadee is indicated by his name, but his song notes of two or three clear whistles, rather long drawn out, are very pleasing in quality.

CAROLINA CHICKADEE.

17   CHICKADEES.

Bird Studies.

The Carolina Chickadee is a bird very closely related to the Chickadee
and is its Southern prototype, occupying the region from middle New Jersey
and Illinois south. It is a much smaller bird, rarely ex-
Carolina Chickadee. ceeding four and a half inches in length. Its colors are
Parus carolinensis Aud.
similar to those of the Chickadee, but the edging of
white to the wing and tail feathers is less marked and there are *no white
edges to the larger feathers of the shoulders.*

The nesting habits and eggs are similar to those of the Chickadee; the
song is weaker and more broken. Though seen about houses, this is decidedly more of a wood bird. The food is almost entirely of insects.

The Hudsonian Chickadee is another close ally of the " Blackcap,"
being its more hardy relation of the North. It is also similar in general appearance, but the top of the head is *seal brown*, suffused
Hudsonian   with gray instead of *black*. The ash of the back has a
Chickadee.
Parus hudsonicus Forst. distinctly brown tinge. The wings and tail are gray,
rather than ash. The throat is black. The sides of the
head and neck, the breast and belly, are white. The sides of the body are
chestnut brown.

Its nesting habits, eggs, and general economy are very like those of the
" Blackcap," but its notes are quite different. It is found from Northern
New England and Michigan northward.

A shower of music, a burst of melody, a thrill of ecstasy, this is the
song of the House Wren. You will hear it about New York during the
third week in April, and from that time on till late July it
House Wren.   is continuous. The musician is a small, alert, inquisitive,
Troglodytes aëdon Vieill.
and aggressive bird. His home for the summer is care-
fully chosen, and when once selected is defended from every intrusion. It
may be a cranny in an outbuilding, a hollow in an old apple tree, a bird box,
or a crevice or knothole in a fence rail, or some nook or corner, perhaps on
your very porch itself. The nest is of twigs, grasses, and feathers, completely
filling the hollow, and entailing more or less labor, on the part of its small
occupant, corresponding to the size of the interior. From five to eight pinkish brown eggs are laid, sometimes unmarked, but frequently finely specked

with a darker shade of the same color. They are more than two thirds of an inch long, and about half an inch broad.

The bird is dull brown in color on its upper parts, becoming brighter on the rump and tail. The feathers on the rump have concealed whitish spots and are of a downy character. The back and rump usually are distinctly barred, but sometimes lack such markings. The throat and under parts are grayish white, and the sides and flanks are more or less barred with blackish brown. The wings and tail are finely barred with blackish brown. The sexes are alike and the young not dissimilar in appearance. The length of the bird is about five inches. They are migratory and are distributed throughout Eastern North America, breeding as far north as Manitoba, Southern Canada, and Maine. They winter in the southern part of their range.

HOUSE WREN. ABOUT THREE WEEKS OLD.

Western House Wren.
Troglodytes aëdon aztecus Baird.

This bird is a close ally and representative, taking the place of the House Wren of the Eastern States, in Illinois and Minnesota. It is a bird similar in size and habits, but is more grayish brown, and generally lighter in color in its upper parts. The barring of the back and rump is more distinct, the color of the bars being blacker. The range of this bird is throughout the Western United States, except those of the Pacific Coast.

Bewick's Wren.
Thryothorus bewickii(Aud.).

In parts of the range of the House Wren, particularly in the South, west of the Alleghanies, there seem to be localities where that bird's place is taken by a wren of about the same size, but slimmer and with a decidedly longer tail and bill. This, too, is a house bird, and except in unsettled districts selects building places similar to those of the House Wren. At a point in the Alleghanies, not far from Roanoke, Virginia, known as Mountain Lake, I found these birds breeding commonly, and there appeared to be no House Wrens proper in the region. Just in front of the large hotel, one of the hitching-posts had been bored with an auger. This hole had been somewhat enlarged in the heart of the post by decay, and here a pair of Bewick's Wrens nested. The song was clear and bolder than that of the House Wren and seemed to have more the air of a conscious performance.

The upper parts of Bewick's Wren are dark umber brown. The rump has the same concealed white spots noticed in the House Wren. The larger wing feathers are unmarked. The two central tail feathers are barred, and the outer ones are blackish, with grayish white tips. There is a grayish white line on the side of the head, over the eye. The under parts are gray and the flanks brown. The bird is about five inches and a quarter long.

It ranges in Eastern North America chiefly west of the Alleghany Mountains, being rare locally east of the mountains. It is found as far north as Maryland and Delaware, and extends irregularly north in the Mississippi Valley to Southern Minnesota. It ranges west as far as the eastern edge of the Plains. It winters in the southern portion of its range. The nest is very like that of the House Wren. The eggs are four to seven in number, and are pinkish white when fresh, spotted with brown and lavender, generally more thickly at the larger end. The eggs are nearly seven tenths of an inch long, and about half an inch in their other diameter.

This is the largest wren that we have in Eastern North America. A bird of the more southern regions and generally resident where it occurs, it

Carolina Wren.
Thryothorus ludovicianus
(Lath.).

is found from the Gulf States north to Southern Connecticut and Northern Illinois. In the region about Washington and to the south it is a common and conspicuous bird, both on account of size and song. The sexes are much alike, and the young on leaving the nest are similar to the old birds in appearance.

The birds are rusty brown above, varying in shade in different individuals. This color is duller on the top of the head and becomes gradually brighter, being brightest on the rump, which has concealed downy white spots. There is a stripe above the eye on the side of the head of white or buff. This is divided from the crown by a narrow black line, sometimes obscure. The wings and tail, rather darker than the back, are finely barred with dark brown or black. The lower parts vary, from dull gray white to deep buff, the throat usually being lighter. The feathers below the tail are barred with dusky markings, and sometimes the flanks, too, are barred with blackish. The length of the bird is rather more than five inches and a half.

The nesting economy is not unlike that of a House Wren, and frequently the nests are found about buildings in nooks and crevices.

Near my home in Florida, a pair one year built a nest on a beam of an outhouse, entering through a knothole. But the woods are more attractive to these birds; holes in old trees and stumps are favorite sites for nests. The nest is a large structure for so small a bird, is made of grasses, leaves and feathers, and lined with fine material of a similar character. The eggs are four to six or seven in number, rosy white, thickly sprinkled with reddish

CAROLINA WREN.

brown and lavender markings. They are three quarters of an inch long and nearly three fifths of an inch around. The nervous energy of the wren family is emphasized in these birds. Never at rest during waking hours, they peer into every nook and corner, dodging from one point to another, in sight for a second, gone before fairly seen, and appearing at another point least anticipated, for an equally brief moment. Like Bewick's Wren the Carolina Wren seems to pose when he sings, as if aware of his musical attainments, and the song is at once noticeable for its fine ringing clearness and an admixture of round whistling notes that distinguish it. It is one of the loudest and most pronounced of small birds' songs.

CATBIRD.

This near relative to the foregoing bird, inhabiting South Florida, is larger than the Carolina Wren, reaching six and a quarter inches in length.

Florida Wren.
Thryothorus ludovicianus
miamensis Ridgw.

It is deep chestnut brown above, and tawny brown beneath, paler on the belly. It is more often barred on the flanks with blackish brown than its congener. In song and general economy it does not differ from the Carolina Wren. But it seems more a bird of the woods than its near relative.

The Catbird is a prying, nervous, wren-like bird nearly nine inches long, clad in a lead-colored coat, relieved by a black cap on the top of the head, and with the feathers below the tail bright rusty brown.

Catbird.
Galeoscoptes carolinensis
(Linn.).

The top of the tail, the bill, and feet are black, and the eyes brown. This is one of our finest songsters, rivaling in varied and brilliant attainments his near relatives the Mockingbird and the Brown Thrasher. Low trees and bushes in the vicinity of the house afford a favorite nesting place. The structure is rather bulky and loosely made of leaves, twigs, and grasses lined with finer grasses and rootlets. The unspotted eggs are of blue green, more than nine tenths of an inch long, and nearly seven tenths of an inch wide. The young when ready to leave the nest are much like the parent birds, save that the black of the head and tail, and the brown feathers beneath it, are duller and fainter. There is a tendency to obscure, transverse mottling on the lower parts, and the back is suffused with brown. If you would see devoted parents, watch the pair of Catbirds that breed in the lilac bushes near your house. Observe the care and solicitude shown their young, with what tenderness they are watched and guarded, and finally how well and judiciously they are trained and educated. The Catbird is found in the Eastern United States and the British Provinces, west to and including the Rocky Mountains ; occasionally on the Pacific Coast, from British Columbia south to Central California. They breed from the Gulf States north to the Saskatchewan.

The Mockingbird is so familiar to most of us, either in his native state or as a contented captive, as to need only a brief description.

Mockingbird.
Mimus polyglottos (Linn.).

He is ashy gray above, with wings and tail blackish and a patch of white on the edge of the wing at the base of its longer feathers. He is light gray or dirty white beneath, and the whole of

CATBIRD'S NEST.

the outer tail feather, most of the next, and half of the third is white. The eyes are yellowish green or grayish white. The young are similar to the adults and both sexes are alike. The birds are from ten to eleven inches long. They are found throughout the Southern United States, where they are resident. From Maryland northward to Massachusetts, they are rare or casual. In the interior they are found regularly as far north as Southern Ohio and Colorado. They breed throughout their range.

In the South the orange trees form favorite places for nesting, and water oaks and bushes particularly about houses are often chosen to build in. A rather loose nest of coarse twigs and rootlets lined with fine grasses, cotton, feathers, and the like contains from three to six pale greenish blue eggs, sometimes almost white or pale buff in ground color, heavily spotted with

MOCKINGBIRD.

reddish brown at the larger end. They are nearly an inch long, and a little less than three quarters of an inch in the other diameter.

So much has been written in celebration of this songster's voice, that to eulogize him further is beyond my power. The song is wonderfully varied, and the musician is tireless, often singing throughout the moonlight night and only resting during the more heated hours of sunlight. Like the Catbird the Mockingbird shows sympathy with his kind. One that I reared in Arizona, when six weeks old assumed the care of two other very young Mockingbirds and a young Oriole, that were placed in the cage with him.

He showed them how to kill and tear apart the grasshopper placed in the cage for food, how to moisten the fragments in the water cup, and generally fed the small birds and looked after them before attending to his own wants. He grew very tame, was allowed the freedom of the room, and would often leave his cage in the evening, after the lamps were lighted, to catch the different insects attracted through the open windows by the light. He would hop about the table at which we were sitting and catch such prey as suited him best. He began to sing softly but very charmingly when about six months old.

Early in May the flood-tide of migrants returning from the South to the points where they breed, is swelled by the Wood Warblers. Of these, conspicuous in our gardens and orchards is the Black-poll Warbler. A leisurely bird in its movements and occupying a very considerable time in passing by to its northern breeding-place, it is to be seen in numbers after the first week in May, until the beginning of the second week in June.

Black-poll Warbler.

Dendroica striata (Forst.).

The adult male bird has a black cap and white sides to his head, reminding one of a Chickadee. The back and rump are gray, streaked with black. The under parts are white, becoming grayish on the flanks and sides, which are streaked with black. The throat is bounded by black streaks, converging to the chin. There are two white bars on the wing. The inner side of the outer tail feathers have more or less white at the tips. The female is quite different, being olive green, streaked with dusky above and having the under parts yellowish white, with sides streaked obscurely with dusky. The wings and tail are similar to those of the male bird. In the fall the adults of both sexes are similar to the females in spring, but are brighter olive green above and below. The striping below on the sides is very obscure if at all apparent. The immature birds of both sexes are brighter olive green above than the adult female, the streaks are indistinct, the lower parts are brighter olive green with the streaks almost obscured. *The feathers below the tail are white.* The birds are about five inches and a half long.

These birds are found in Eastern and Northern North America. They breed from Northern New England, north to Greenland and Alaska. They winter in Northern South America.

They breed in coniferous trees, generally not more than eight feet from the ground. The nest is of moss and rootlets, having a lining of fine grasses.

About the House.

The eggs are from three to five in number, white in color, spotted heavily with reddish brown, particularly on the larger end. They are almost three quarters of an inch long, and a little more than half an inch in width.

Another of the wood birds seen about our houses is the Myrtle Bird or Yellow-rumped Warbler. These birds are so definitely marked in the spring as to be easily identified. Old males are conspicuous by **Myrtle Warbler.** the bright yellow patch on the crown and rump, supple-
Dendroica coronata (Linn.). mented with a patch of the same color on each side of the breast, in strong contrast to an area of intense black between them. The throat is white. The upper parts are bluish gray, streaked with black. The outer tail feathers have white spots on their inner margin, near their tips. The adult female is similar to the male, but much duller in color. In the autumn and winter the sexes are still more alike. There is a general suffusion of brown, darkest on the upper parts, obscuring the brilliant pattern of colors, which still may be detected, but more plainly if the plumage is ruffled. Birds in first plumage lack the yellow, except that some individuals have it on the rump. The whole plumage is thickly streaked with dusky brown and grayish white. The length of the bird is five and three quarters to six inches.

The nesting habits are similar to those of the Black-poll Warbler, and the eggs, from three to five in number, are white, spotted rather obscurely, especially at the larger end, with reddish brown or olive brown. They are about seven tenths of an inch long and rather more than half an inch broad.

The birds are found in North America but chiefly east of the Rocky Mountains. They breed from the Northern United States northward, and are found in winter from Southern New England through the West Indies and Mexico to Panama.

Audubon's Warbler is a bird of Western North America, which ranges east to the eastern base of the Rocky Mountains. It has been recorded once from Pennsylvania and also from Massachusetts as
Audubon's War- an accidental straggler.
bler.
Dendroica auduboni In general appearance and color it closely resembles
(Towns.). the Myrtle Warbler, and is of about the same size, but has a *bright yellow, instead of a white throat.*

Its breeding habits are very much like those of the Myrtle Warbler.

Bird Studies.

When you see a bird that is bright yellow all over, brighter on the crown, and with decided reddish streaks on the under parts, it is an adult male Yellow Warbler. The female is similar to the male, but **Yellow Warbler.** the under parts are often plain, with very slight reddish Dendroica æstiva (Gmel.). markings. The young are similar, but the yellow is duskier and there are no markings below. *In any plumage* a Yellow Warbler may be known by *the yellow of the inner webs of the tail feathers.* The birds are rather more than five inches long.

There are two or three Wood Warblers that are commonly called "Wild Canaries." This is one of them and probably the bird

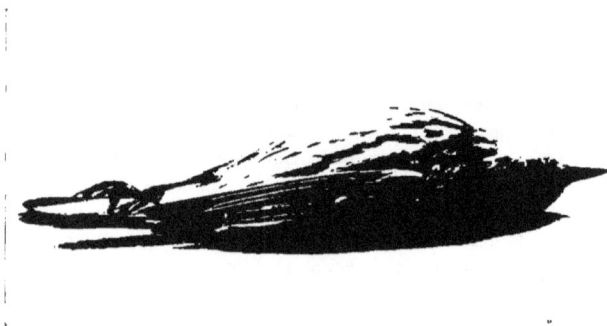

YELLOW WARBLER. ADULT MALE.

most generally meant in such connection. The others will be dealt with as we meet them. Suffice to say, there are no "Wild Canaries" in North America, the true Canary being a Finch or Sparrow in its affinities, while the Yellow Warbler is one of the so-called "soft-billed" birds belonging to a family widely separated from the Finches. Yellow Warblers find nesting sites in the trees and bushes about houses, and also along the banks of ponds and streams and the edges of swamps. Here in some suitable crotch or fork, generally not more than fifteen feet from the ground, a nest of fine grass, woven with fibres of plant stalks and fern down, is built. The lining is of finer material of a like character, and horse hairs are generally a feature of the interior. The eggs are from three to five in number, greenish white in color, spotted with varying shades of brown, reddish, olive, and lilac. These spots form a sort of wreath of darker color near the larger end of the egg. The eggs are about two thirds of an inch long and a little less than half an inch broad.

This is one of the smaller birds persecuted by the Cowbird, the eggs of which are often placed in the Yellow Warbler's nest.

The birds are found in North America, except the southwestern portion. They breed practically throughout this territory, and winter in Central Amer-

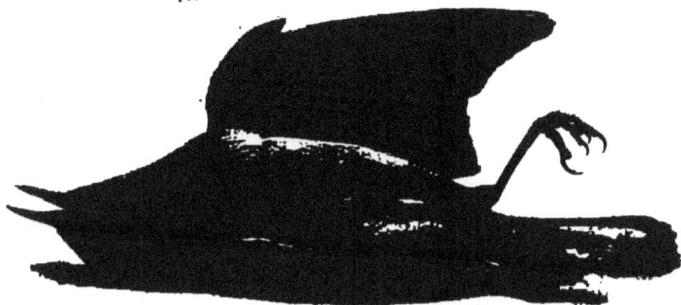

YELLOW WARBLER. YOUNG MALE.

ica and Northern South America. Their food is almost exclusively insects, and like the other Warblers they benefit the farmer.

One of our most persistent song birds arriving during the last week in April or early in May, is the Red-eyed Vireo. The birds are very common about houses and in the woodlands, in fact wherever there are trees. They remain in the region near New York till well into October.

Red-eyed Vireo.
Vireo olivaceus (Linn).

A bird about six inches and a quarter long ; it is to be known most readily by its head markings. The slate colored crown, with its border on either side of black or very dark slate, the well marked white line above the eye and next to the blackish border of the crown, are sufficient alone to iden-tify it. The general color of the remaining upper parts, wings and tail included, is olive green light in tone and of a grayish tinge. The under parts are pure white, except that in some individuals the feathers below the tail are yellowish white. This feature is particularly characteristic of birds taken in the fall. The iris is generally red with a carmine tinge, but some individuals in spring and many in fall have brown eyes. The sexes are alike, and *the young* are very similar to the old birds, *but always have brown eyes.*

The nest is pensile or semi-pensile and is composed of grasses, strips of plant fibre, and plant down. These materials are woven compactly and smoothly, and the nest is lined with a finer coating of grasses and bark-fibres. The upper edge of the nest, for about half its circumference, is attached to a point where a small branch forks, generally toward the termination of a limb. The height from the ground varies from five to forty feet, rarely higher. Such nests are to be found in trees along busy streets of country towns, about our houses and in the thicket and woodland. The eggs are

RED-EYED VIREO.

white, with a slight freckling of black or brown dots on the larger end. They are a little more than four fifths of an inch long, and nearly three fifths of an inch in their other diameter. All through the long summer days, even in the noon-day heat of July and August, the Red-eyed Vireo sings—a slow, drowsy, broken song. Hesitating as if at a loss for the next series of notes, the pause is long, but they are sure to come. It seems as if he waited to hear some other birds that have long since ceased to fill their part in the general chorus.

The bird finds a winter home in Florida, and from there to Northern South America. In summer it ranges throughout Eastern North America, north to the Arctic Circle and west to British Columbia. It breeds in almost its entire North American range.

NEST OF RED-EYED VIREO. AFTER YOUNG HAD FLOWN.

This bird is about four inches and a half long. It is plain dusky slate color above shading into black on the top of the head, and with a yellow **Bahama Honey** patch on the rump. There is a clear stripe of white above **Creeper.** the eye. The cheeks are white and the region above the Cœreba bahamensis (Reich.). ears black. The throat is white, reaching well down on the chest. The breast is yellow and the sides and flanks grayish tinged with yellow. There is a conspicuous white patch at the base of the larger feathers of the wings. The feathers below the tail are white and some of the outer of the dark tail feathers are tipped with the same color.

The birds build oven-shaped nests with an opening on one side. They are placed in low trees or bushes. The eggs are white, sprinkled with fine dark brown specks. They are about two thirds of an inch long and half an inch broad. This is a bird of the Bahama Islands and also occurs on the keys, and probably the mainland of Southern Florida.

The Waxwings are the nomads among birds, the gypsies in feathers, the wanderers without regard to season. Their marked traits are gregariousness, great personal beauty in dress, a certain uncan-**Cedar Waxwing.** niness and mystery in their silent comings and goings, and Ampelis cedrorun (Vieill.). a love for forbidden fruits. No song nor season heralds their appearance. The last flock I saw were feeding on some overlooked cedar berries, when the ground was white with snow, in January; and when the cherry has taken the place of its blossoms, on some warm day in June, offering the greatest contrast to the wind-swept snow of our last meeting, I shall see the gypsies again. They will be as much at home in the warm sunshine of June, when nature is at its flood, as they were in the cedars and snow, when the hand of the ice king had wrapped sleeping life in his cold white mantle.

How delicately the soft brown of the back shades into the gray blue on the rump and upper tail! What a fine gold tip has each tail feather! The velvet black of the chin and about the eyes is in fine contrast to the brown crest, that lends so much character to their appearance. Where the soft jet black of the chin changes into brown on the throat and breast no one can tell. The same subtle kind of change into the faint but decided olive on the flanks, and to white on the belly and the feathers below the tail, leaves no abrupt marking or outline that can be defined. Many of the birds have very fine coral red tips to some of the smaller wing feathers, and now and then an individual has the gold end of each tail feather tipped with these brilliant decorations.

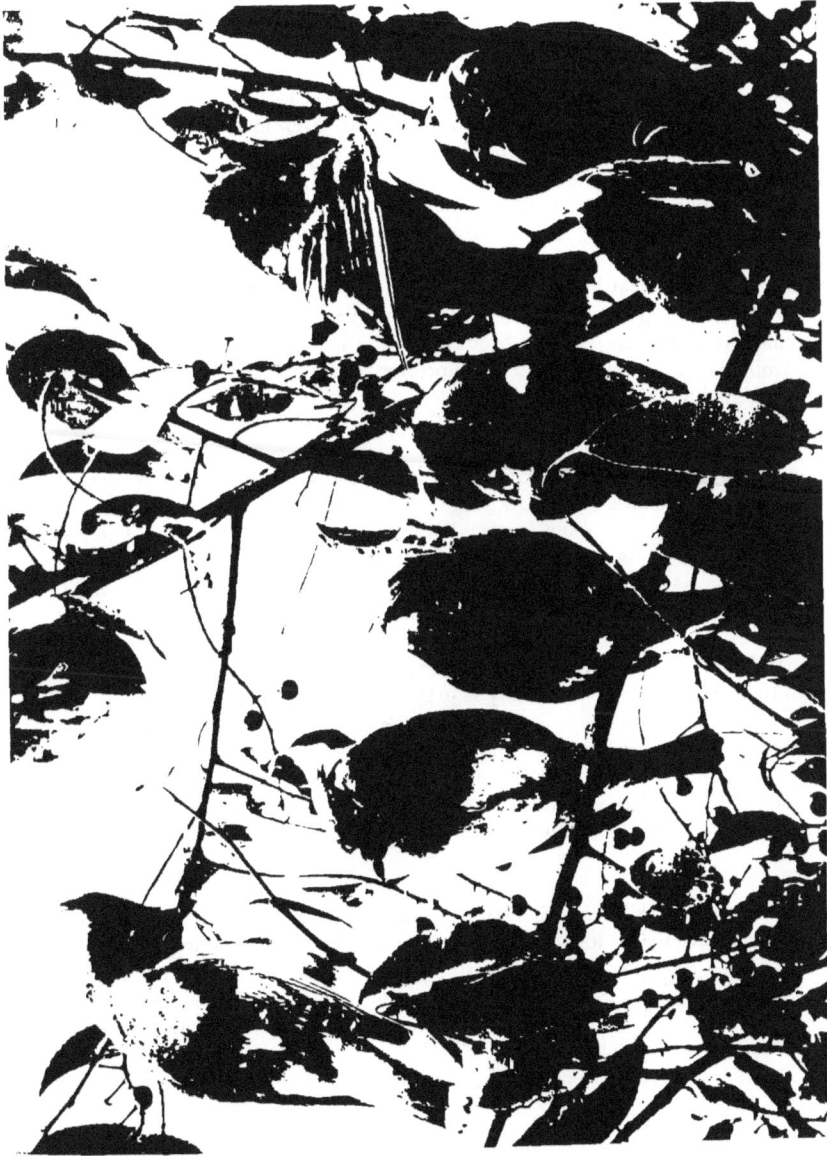

CEDAR BIRDS.

33

The sexes are alike, and the young on leaving the nest are duller than the old birds and have the lower parts streaked with brownish gray on an obscure white ground. The birds are about seven inches long.

Late in June or during the first week in July, a bulky nest is built in some tree, generally near the house. It is made of grass, strips of bark, moss, and sometimes mud, and is lined with fine grasses. From three to five pale purplish gray eggs are laid. These are marked, with a varying degree of distinctness, in spots of black and brown. They are about nine tenths of an inch long and somewhat more than three fifths of an inch broad.

The birds are found throughout North America; they breed from Virginia northward, and in the altitudes of the Alleghanies farther south to the Carolinas. They winter from the northern border of the United States to Northern South America.

This is a bird like the Cedarbird in general appearance, but somewhat larger, being about eight inches long. It has the same velvety markings on **Bohemian Wax-** chin and face, and even a more conspicuous crest. There is **wing.** a decidedly reddish or deep cinnamon tinge about the Ampelis garrulus Linn. brown of the head, and the feathers below the tail are chestnut red. The feathers of the shoulders are tipped with white. The coral red wax tips to the wing feathers, and often to the tail feathers, are very conspicuous. The longer wing feathers are tipped with white or yellowish on their outer edges. The belly is gray.

The nest and eggs are similar to those of the Cedarbird, the eggs being larger. The Bohemian Waxwing is almost entirely confined to the northern parts of the Northern Hemisphere. It has been recorded as a rare or accidental visitor in the Eastern States, and with more or less regularity and frequency in the upper Mississippi Valley.

Early in April come the Barn Swallows, graceful and fleet, true aërial hunters, now flying low over field and meadow, now skimming the placid **Barn Swallow.** surface of pond or stream. They are veritable house Chelidon erythrogastra (Bodd.). birds, making their nests and rearing their young with our chickens and cattle, almost as tame and quite as much at home.

This is the Swallow with the deeply forked tail, with the upper parts, except the forehead, steel blue, the forehead and throat being bright chestnut. This color becomes paler on the breast and belly, particularly in the female. All the tail feathers but the two middle ones have white on their inner webs. The adult birds are almost seven inches long. Immature birds are pale below and dusky above, the outer tail feathers are shorter, making a much less forked tail.

BARN SWALLOW.

The birds are gregarious, generally several, and often many pairs nesting in the same barn, where the beams and rafters afford shelves on which they place their nests. These structures are open, built of mud pellets mixed with straws and grasses, and have a lining of finer grasses and feathers, often those shed from the poultry in the yard hard by. They are attached by the sides to the wall of some cave or of an outbuilding. Sometimes they are placed on a ledge or beam which answers for a shelf to support them. The eggs, four to six in number, are about three quarters of an inch long and rather more than half an inch in their other diameter, and are white in color, spotted with brown. The birds range throughout the region under consideration, extending through North America to Greenland and Alaska. They winter in Central and South America, and breed in their habitat as far south as Mexico.

Their food is exclusively insects, is generally taken on the wing, but I have seen them, at times, feeding on the bare ground. This is one of the birds that has been noticeably influenced in its general breeding habits by the advent of civilization. Presumably they formerly found nesting sites on the sides and ledges of caves, and there is little doubt that in remote or thinly settled regions this habit still obtains.

A broader, stouter bird than the last is the former Cliff Swallow, now the Eave Swallow of our barns and outbuildings. Both names indicate the **Cliff Swallow.** situations chosen for their nests, but with the advent of Petrochelidon lunifrons houses, the cliffs and caves are all but deserted. In gen-
(Say). eral color the birds are similar to the Barn Swallows, steel blue and bright chestnut prevailing. But the forehead is white or very pale brown, and *the rump is pale brown*, thus readily distinguishing them from the

CLIFF SWALLOW.

Barn Swallows. The rest of the upper parts, including the wings and tail, are dark steel blue, and there is a patch of similar color on the lower part of the throat. *The tail is short.* The birds vary from five and a half to six inches in length. They arrive some ten days or two weeks later than the Barn Swallow and are emphatically gregarious in the breeding season. They build *closed* nests, the walls composed of mud pellets packed together, and a small entrance being left, sometimes at the top, but often on the side of the wall low down. These nests vary greatly in shape, and are placed in the angle formed by the eaves of barn or other building. The lining is sparse, consists of some grass and feathers, and is distinct and separate from the wall of the nest. The eggs vary from three to five in number, are very similar in color to those of the Barn Swallow, but are generally larger. The birds range throughout North America, except Florida, to the tree limit. They breed south to the Potomac and Ohio valleys in the area treated of.

The Cuban Cliff Swallow, similar in general appearance to the last, but smaller, is a native of the West Indies, where it is an abundant bird.

Cuban Cliff Swallow.
Petrochelidon fulva (Vieill.).

The author took two representatives at Garden Key, Dry Tortugas, Florida, in March, 1890, which are its North American records. It must be regarded as an irregular or accidental visitor.

This is the largest and heaviest of our Swallows. A bird eight inches long, the adult male has a uniform of lustrous bluish black, the wings and tail being duller than the body plumage. The female is not as

Purple Martin.
Progne subis (Linn.).

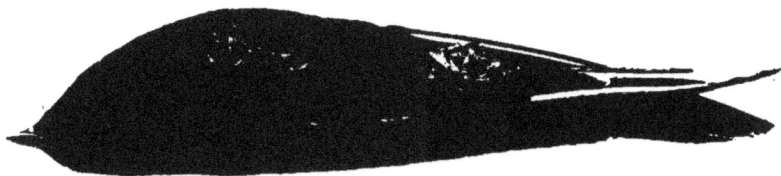

brightly colored above as the male, and the under parts and sides are grayish brown, the feathers being tipped

PURPLE MARTIN.

with white, which color prevails on the belly.　The immature birds resemble the female.

Though the birds range throughout North America as far north as New-foundland, in the breeding season they are only of general occurrence in the more southern parts, being local if not rare in the States north of Delaware. They breed in boxes and other shelters erected for their use about the house and yard.　Here are laid four or five pure white eggs, about an inch long and nearly three quarters of an inch in their smaller diameter.　More musical than the other representatives of the family, their gay carol is a welcome addition to the bird chorus about our houses.

The Cuban Martin is very similar in general appearance to the last, but in the adult male the ventral region shows a spot or bar of white, beneath the surface feathering of bluish black.

Cuban Martin.
Progne cryptoleuca Baird.

It is found in North America in Southern Florida, and its general economy is identical with that of its near relative.

A " Redbird," his home name, best describes this gay fellow, and if it is added that he has a pronounced crest like that of a Jay or a Cedar bird, and that he is between eight and nine inches long, our bird is

Cardinal.
Cardinalis cardinalis (Linn.).

identified.　This is the male : the female is *identical in shape ;* the varying shades of red in the male being re-placed by buff or brownish snuff color, with suggestions of red on the crest, wings, and tail.　Both have the forehead and sides of face *and the upper throat* more or less black or dusky, varying in tone in different represen-tatives.

This bird is not a migrant, and though not as common at its northern limit, Southern New York and the Great Lakes, as it becomes farther south, yet the Cardinals that live in these regions brave the cold and storms as does the equally brilliant Jay.

But it is throughout Virginia and farther south that this is one of the house birds.　Both sexes are brilliant songsters, are heard at all times and are seen almost as frequently.　In Florida they are ubiquitous.　There, daily throughout the winter, I have seen them feed on seeds and crumbs, often five or six together, close to the windows of my house.　Their nests of dead leaves

39 SONG SPARROW'S NEST AND EGGS.

and twigs, with a finer lining of grasses and rootlets, are placed in bushes and thickets, often in low, damp places, and three or four bluish white eggs, spotted with reddish brown and splashed with grayish lavender, are laid. These eggs are about one inch long and a little less than three quarters of an inch in their median diameter.

The geographical race, found throughout Florida, is known as the Florida Cardinal. The birds are smaller and generally darker colored than the Car-

Florida Cardinal. dinal of the North, and the female shows a stronger tone of
Cardinalis cardinalis flori- red in the crest and on the tail. In this sex the breast and
danus Ridgw. chest are often strongly tinged with red, or there is an intermixture of decidedly red feathers.

As the photographs and portraits of those we know best and are most familiar with are generally disappointing, because of our very knowledge and

Song Sparrow. association, so any description of this bird seems inade-
Melospiza fasciata (Gmel.). quate, careful as it may be.

Shall I say that the Song Sparrow is a brown bird, streaked with darker shades of brown above ; that beneath it is white, the sides and breast streaked with brown, and the white throat clearly defined by a brown stripe on either side ? That there is a more or less defined brown spot on the striped breast ? That he is about six and a half inches long ? Do you recognize this pen picture of an old friend ? If not, does the portrait opposite recall the minstrel whose happy song begins when the ground is white with snow, and closes with the last of the falling leaves ?

With us Song Sparrows are resident and present throughout the year. They breed from Virginia and Illinois north to the Fur Countries, and winter from Massachusetts to the Gulf States.

· The nest is a rather loose and bulky structure composed of leaves, coarse grass, and strips of bark, lined with similar material of a finer character. It is generally placed on the ground, but now and then in low bushes. The eggs are three to five in number, bluish white in color, splashed heavily with brown, so as to almost hide the ground color. They are nearly four fifths of an inch long and almost three fifths of an inch broad.

SONG SPARROW.

Bird Studies.

This is our gregarious winter friend, in slate blue coat and hood, with a color below that matches the snow. When the Snowbird flies, the tail is

Slate-colored Snowbird.

Junco hyemalis (Linn.).

slightly spread and a pure white border is added, by the color of the outer tail feathers, to the tail of dark slate blue. The female is paler in color. Many individuals, especially young birds of the year, have a suffusion of brown over the slate color. Very young birds in their first plumage have the entire lower parts grayish white, streaked with dusky black, and the upper parts obscure slate color, streaked with black. This bird is about the size of a Song Sparrow and of very similar build.

SLATE-COLORED SNOWBIRD.

The nest is placed on the ground, often where a tree has blown down, among the exposed roots, and again on some bank along a road where the wash of rains has left an overhanging turf. It is a loose structure of moss, grasses, and rootlets, lined with hair and fine grass. The three to five eggs are white with specks of reddish brown, often becoming splashes at the larger end. They are three quarters of an inch long and five eighths of an inch in their other diameter.

This is a bird of Northern North America breeding from Northern New York, northward, and southward on the higher Alleghanies to Virginia. In winter it is common throughout the Eastern United States, as far south as Georgia.

The Carolina Snowbird is very similar to the last, and so closely allied as to be regarded as a geographical race. It inhabits the higher mountains of Virginia and North and South Carolina, where it is com-

Carolina Snowbird.
Junco hyemalis carolinensis
Brewst.

mon and breeds in numbers. It is rather larger and paler than our Snowbird. The head is colored like the rest of the upper parts. The brownish suffusion in the female and immature birds is not so apparent as in the Snowbird. Its method of nidification is identical with that of the common Snowbird, and the eggs are similar.

This bird is a western form, but has been taken in Illinois, Maryland, the District of Columbia, and Massachusetts. It is very

Hybrid Snowbird.
Junco hyemalis connectens
Coues.

like its near relative our Snowbird, but has a browner back, with the sides decidedly brown or pinkish brown of a lighter shade.

One of the most sociable birds about our houses and grounds during the warmer portions of the year is the Chipping Sparrow. Its nesting sites

Chipping Sparrow.
Spizella socialis (Wils.).

are the currant bushes and other shrubbery of our grounds. The vines on the piazza are an invitation to sociability and perhaps protec- tion, which it often accepts in locating his little home and rearing its family. As the bird is so well known to us,

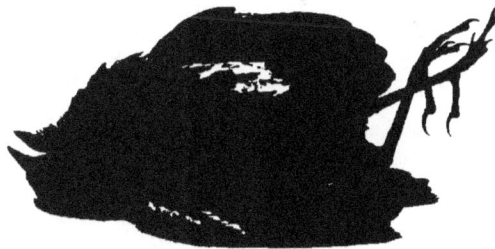

CHIPPING SPARROW. ABOUT THREE WEEKS OLD.

so perhaps the nest is, of small nests, the most familiar. Round and rather shallow, it is

CHIPPING SPARROW. ADULT MALE.

made of grasses and small twigs lined noticeably with long horse hairs. Generally located near the ground, it is often found ten to twenty feet above it. The eggs are blue green, dotted and marked with red and dark brown or black. They are three to five in number, rather less than three quarters of an inch long, and half an inch in their smallest diameter.

The salient characteristics of the adult birds are, a *black bill* and forehead, a *chestnut brown crown*, and *gray rump*. *The under parts are of ashy gray*, the throat whiter. Very young birds have the breast streaked or spotted with dark brown, almost black, have a lighter colored bill, lack the forehead markings and chestnut crown. They soon pass from this stage to the second plumage, which lacks the breast markings, but is otherwise similar to the first stage. In all phases of plumage the ashy color of the rump will serve as a great aid in identifying the bird.

The Chipping Sparrow ranges west to the Rocky Mountains, north to the Great Slave Lake. It breeds from the Gulf States throughout its range, and winters in its more southern breeding grounds.

A yellow bird, with black wings, a bar of white across them, a black cap on his head, and a black tail when closed, but showing the white inner webs **American Gold-** when spread, the rest bright lemon yellow,—such is the **finch.** Goldfinch.

Spinus tristis (Linn.). This is the plumage of the old male in summer. The female at the same season is much duller; the wings and tail are dusky, the black cap above is lacking, the yellow of the back and head is suffused with brownish, and below, the ground color, obscure white, is washed with buffy and brown tinged with yellow, especially on the throat. In winter, the sexes are more alike, the male having lost the black cap and being generally obscured with brownish over the yellow, but still having the tail and wings black, though not so bright as in summer. The young resemble the female bird.

The American Goldfinch is a small bird averaging rather more than five inches long, and has a very characteristic flight, wave like and undulating.

These birds are eminently gregarious, going in flocks of varying sizes, except during the breeding season. Thistles and sunflowers, and in fact almost any seed plant, attract them and furnish their food supply. They breed rather later than most of our birds, building a bulky nest of strips of

bark and grasses with sometimes moss added, the whole being felted with
thistledown and placed in a fork in a bush or tree from six to twenty-five feet

AMERICAN GOLDFINCH. ADULT MALE.

from the ground. The eggs are small, being five eighths of an inch long and
less than half an inch in width. They are pale bluish white in color and
unmarked.

The birds have a wide range through North America, winter mainly in
the United States, and breed from South Carolina to Labrador in the area
under consideration.

This is a European species that has been introduced in the vicinity of
New York and Boston. It is not uncommon locally in and near both cities.
European Gold- Rather larger than our Goldfinch, it may be readily recog-
finch. nized by its bright red face, and black wings with a bright
Carduelis carduelis (Linn.). yellow band or bar. Its habits are described as similar to
those of our bird.

European House Sparrow.
Passer domesticus (Linn.).
This is another European species which, since its introduction about forty years ago, has become one of the commonest birds about our houses, in cities, towns, and villages, and now is too familar to all to need a detailed description or further notice here.

European Tree Sparrow.
Passer montanus (Linn.).
The European Tree Sparrow has been introduced and become naturalized in the city of St. Louis, Missouri, and vicinity. Its habits are very similar to those of its more widely distributed ally, which it resembles in general appearance.

Purple Finch.
Carpodacus purpureus (Gmel.).
A rose colored sparrow seems an anomaly, but such is the adult male Purple Finch. Individuals vary greatly in intensity of color, but a typical male has a predominant rosy red color which is browner on the back, wings, and tail and which shades into white on the belly. The females are quite different, they lack all traces of the reddish color, and have the general appearance of a striped sparrow, the prevailing colors being olive and gray on a grayish white ground, becoming clear white on the under parts. The young birds of both sexes resemble the adult female, but lack much of the olive shading, the gray being darker and on the back tinged with brownish.

The birds are rather more than six inches long, have a heavy, conical bill, with pronounced feather tufts over the nostrils and a distinctly forked tail.

The nest is a shallow, thin affair, placed on some horizontal limb, generally in evergreen trees, at varying distance from the ground, but rarely above thirty feet. It is built of roots and grasses and lined with finer material and long hairs. The eggs vary from three to six in number, are blue in color, spotted at the larger end with brown, and are four fifths of an inch long and less than three fifths in their other diameter.

A bird of Eastern North America, it breeds from Long Island north and winters from the New England States south to the Gulf. At many places where it breeds it is familiar in gardens and about houses. Its food is seeds and berries, juniper and the like, young leaf buds and the blossom buds of many fruit trees.

PURPLE FINCH.

About the House.

There are three kinds of Crow Blackbirds in the Eastern United States. They are large, conspicuous birds, black in color with varying iridescent shades of metallic lustre on their coat. Their eyes are yellow. They are closely allied, and, grading into each other, are not specifically distinct, but form three geographical races. They vary in size from twelve to rather more than thirteen inches in length. The females are rather duller than the males, lacking much of their iridescent brilliancy. They breed in communities, often in pines or other evergreen trees, well up from the ground, but sometimes their nests are placed in bushes or in holes in trees.

The nest is a large structure, built of mud and grass and lined with finer grasses. From three to six bluish green eggs are laid, which are spotted and marked with zigzag lines of varying shades of brown.

In the lower Mississippi Valley and east of the Alleghanies from Georgia to Massachusetts, the Purple Grackle is the representative of the trio in
Purple Grackle. the breeding season. The general tone of the male bird
Quiscalus quiscula (Linn.). is lustrous purple with bluish green or steel blue sheen. The rump is greenish purple or clear purple and the feathers of this region show *defined iridescent barring.*

The smallest of these three races of Crow Blackbirds is the Florida Grackle. It is found in Florida and the southern part of the Gulf States to
Florida Grackle. Texas. On the Atlantic coast it ranges as far north as
Quiscalus quiscula aglæus the Carolinas. In this race, the entire head is metallic
(Baird). purple with a violet sheen, which color extends back on the neck and above to the breast below. The general color of the back and rump is iridescent green, the latter region *showing traces of iridescent barring. The feathers of the shoulders frequently have iridescent tips.*

The Bronzed Grackle is nearly the same size as the Purple Grackle. East of the Alleghany Mountains, it breeds from Southern New England to
Bronzed Grackle. Newfoundland, west of the Alleghanies from Texas to
Quiscalus quiscula æneus Great Slave Lake.
(Ridgw.). The head and back colors are much the same as in the Purple Grackle, but the body color is bronze with a metallic sheen and *no iridescent bars at any point.*

30   ORCHARD ORIOLE. ADULT MALE.

About the House.

The Orchard Oriole is a bright chestnut colored bird, with black beginning at the region just between the shoulders and extending over the entire head and throat. The wings and tail are almost black, the **Orchard Oriole.** larger feathers having light edges, often dull white. There Icterus spurius (Linn.). is a broad chestnut bar on the wing, and the lower back, belly, sides, and breast are chestnut. Such is the adult male, very unlike his brighter cousin the Baltimore, but the female and young males are nearer oriole color, being greenish yellow above and brighter lemon yellow below, with wings of obscure brownish color, showing two whitish bars. The tail is bright yellowish green. The young males have the back browner and in the second year acquire a black throat patch. Many individuals in this phase of plumage have traces of the ultimate chestnut body color mixed to a varying extent with the yellow and olive green feathers. In the third year the full plumage is generally attained ; still, some individuals show yellow mixed with the chestnut of the body and black of the head and upper back.

They build pensile or semi-pensile nests of fresh green grasses very beautifully and skilfully woven, sewn and bound to the surrounding twigs and leaves. These nests are at first so near the color of their leaf environment as to be inconspicuous, but later turn yellow or dry grass color. They are among the most beautiful and elaborate structures of our native birds, and are situated in an apple, pear, or maple,—in fact, in almost any of the smaller trees about houses,—from three to twenty feet from the ground. The nests contain from three to five eggs, bluish white in color, spotted, and marked in zigzag lines with dark brown and black. They are about four fifths of an inch long and less than two thirds of an inch in their other diameter.

The birds inhabit Eastern North America in the breeding season, from the Gulf States to Southern New England, Michigan, and Ontario. They winter in Central America. This is one of our finest song birds, a foe to insects, and not confined to orchards, but often seen about the garden and grounds close to our door.

Bullock's Oriole is a Western species ranging over the United States, from the Rocky Mountains westward. It goes as far north as Manitoba and British Columbia, and winters from Mexico southward. **Bullock's Oriole.** There is a single record of its capture at Bangor, Maine. Icterus bullocki (Swains.). Its general color is deep orange, which prevails on the forehead and sides of the head, extending down on the neck, dividing the

ORCHARD ORIOLE. ADULT MALE.

ORCHARD ORIOLE'S NEST.

black of the throat from that of the top of the head and back. The rump
and lower parts except the throat are orange. The *shoulders are black*, and
there is a broad white patch on the wing. The tail is orange and black.
The female is much like the female of the Baltimore Oriole in general
appearance, but is noticeably larger. The birds are about eight inches and
a half long.

The nesting and breeding are much the same as in the Baltimore Oriole,
and it takes the place of that bird in the West.

Troupial.   The Troupial, a South American bird, was recorded
Icterus icterus (Linn.).   by Audubon as accidental at Charleston, South Carolina.

Some account will now be given by word and picture of how young birds
grow. The nest on the opposite page is that of the Blue Jay. It is made
of coarse twigs and roots lined with finer material, and
Blue Jay.   though firm and strong is rather loose in construction, even
Cyanocitta cristata (Linn.). in the lining. This nest was placed on a horizontal limb
of a beech tree, where it forked, and was further supported by the upright
small branches that grew at right angles to the main limbs. It was about
eighteen feet from the ground, thirty feet from a dwelling, and almost over a
road where there was constant traffic. Hemlocks, beeches, and other large
trees close by afforded deep shade which rarely admitted any sunshine even
on the brightest days.

In the nest at the time the photograph was made there was one young
bird just hatched and three eggs on the point of hatching. One of these was
" pipped," and shows this condition plainly in the picture. This nest was
found at South Orange, New Jersey, in a yard on the Ridgewood Road, a
street on which there are many residences quite close together. The date
on which the photograph was made was June 16, 1897.

For the first six days the four dusky colored chicks were practically
blind and naked. They could raise their heads to receive the food constantly
supplied by their parents, and elevate their bodies at intervals to evacuate
faeces, which were immediately removed from the nest by one of the old birds.
A far more continuous movement was that of their feet, which were opening
and closing on the fine roots and twigs composing the loose lining of the
nest, almost without interruption.

BLUE JAY'S NEST AND EGGS. ONE EGG PIPPED.

One of the birds was placed on a sheet of paper, and photographed on June 22d. Still a naked chick, with eyes scarcely opened, and with the quills of the wings just appearing, he was weak and helpless, being unable on such a surface either to raise his head or to place his feet so as to sustain himself. He was then six days old and his photograph is appended. In a way he is more reptilian than bird-like in appearance.

BLUE JAY. SIX DAYS OLD.

Returned to the nest, none the worse for his few moments' absence, he was photographed again in the same way on June 26th. The bird now ten days old had some pretensions to down, and its growth and that of the wing feathers was marked in the interval. The feet were now able to hold somewhat to the surface of the paper. The eyes were just opened, but hearing and feeling seemed still the principal sense agents.

The next picture shows another brood of Blue Jays taken in the same yard, presumably about eleven days old. They are in the act of clamoring for food.

This photograph was made on June 12th. As yet there is no definition

BLUE JAY. TEN DAYS OLD.

in pattern of feather color on the birds. They are appreciably stronger,

57 BLUE JAYS IN NEST. ABOUT ELEVEN DAYS OLD.

58   BLUE JAYS IN NEST. ABOUT THIRTEEN DAYS OLD.

however, and their eyes though curiously glassy undoubtedly begin to distinguish objects.

Another photograph of this same brood taken on June 14th, two days later, shows the very rapid growth of the feathers, the color pattern of which can be plainly distinguished on the head and wings.

BLUE JAY. TWO WEEKS OLD.

Let us now return to the brood of birds hatched on June 16th.

On June 30th they were beautifully feathered, were very lively, moving about the nest constantly, and would evidently fly and leave their home in a day or two. Two of the young were now removed from the nest to be reared, in order that observations as to their growth might be continuous.

One of the fledglings was at once placed on a small branch, which he readily grasped, and was again photographed.

This bird was just two weeks old. The growth in a period of four days is so marked, has been so rapid, and is so clearly shown that a detailed account of it is unnecessary. The reader has only to refer back to the picture taken on June 26th to see all this personally.

It may be well to state that in this series of pictures the young birds are portrayed as nearly as possible natural size.

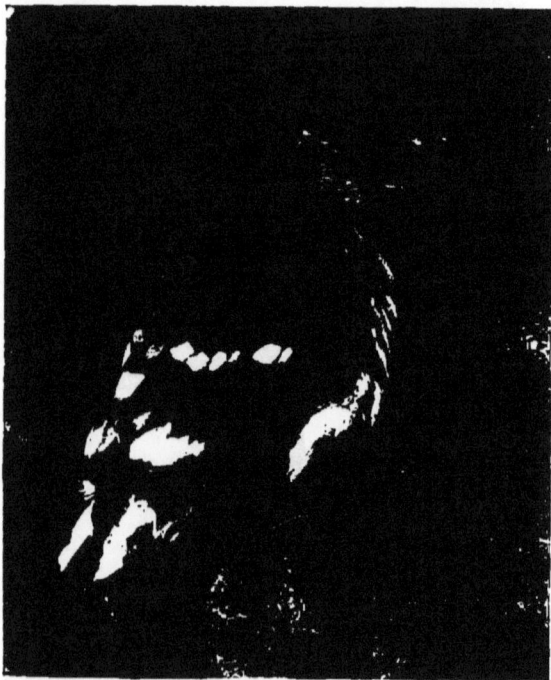

BLUE JAY. SEVENTEEN DAYS OLD.

Two days later, on July 2d, the chicks walked about the artificial nest that had been made for them, and constantly stood erect exercising the wings by flapping them at short intervals.

On the 3d of July they left the nest and could fly from perch to perch

BLUE JAY HAMMERING.

in the large cage that had been provided for them. They were then seventeen days old, and a photograph of one taken at that time is appended.

From this time on they grew rapidly and on July 15th were full grown and fully fledged, looking very like the old birds, save for their somewhat grayer blue feathers and generally softer plumage.

They now began to moult their plumage, except the wing and tail feathers, and during the first week in August had finished this change, having the gray of the back replaced by shiny blue, and the body plumage by clearer colored and stronger feathers. These birds at this writing, October 1, 1897, are still in my possession ; they have undergone some further changes by moulting, all of the feathers of the wings and tail having been replaced, as well as those of the head and throat. They are very tame, and live at liberty in a large airy room with other birds taken in this and previous years, and kept for observation and study.

The reader's attention is called to the fact that the nest of June 16th had only been disturbed when the young birds were taken out for a few moments to be photographed, and that it still remained in its original position. The old birds did not seem in any way to resent the intrusion, but continued the care of their brood. When on June 30th two of the young were finally taken from the nest, it was left to the possession of the parents and the two remaining chicks. It is unnecessary to allude to the solicitude shown young birds by their parents, but too much stress cannot be laid on the extreme care taken in removing all dirt and excrement from the nest, and the resulting cleanliness. This can only be realized by the examination of many recently abandoned nests where broods of singing birds have been raised. Aside from a certain wear and settling they can scarcely be distinguished from those newly built.

A further word in regard to the movements of very young birds seems essential. These movements are accomplished largely by the aid of the feet and legs. At first birds rest on *their entire foot* to the *heel.* They exercise the muscles of their legs by constantly opening and closing their feet. This motion is uninterrupted, save when the birds sleep. How large a factor the *linings* of *nests* form at this period of their lives is only to be appreciated by watching nestlings. As one foot is opened the other closes on the twigs, rootlets, or fine grasses, the bird never slipping and always sustained.

In attempting to raise a brood of Blue Jays, having secured birds about ten days old, I left them in their nest, but thinking to keep the birds clean and free from the dirt they made, I placed a cotton cloth between the birds

63   BLUE JAY.

and the nest.   This was frequently changed in pursuance of my idea.
Three of the nestlings soon lost all use of their legs and feet, which became
bent and deformed, and the birds died in about a week, except one stronger
than the rest, who though arrested in physical development yet managed to
survive.   This result I attribute to the inability of the young ones to
properly exercise their feet and legs on the smooth unnatural surface provided for them, and subsequent experience with Jays, Chats, and other
young birds has confirmed this opinion.   It seems that one of the offices per-
formed by the nest lining is to afford grasping material on which very young
birds begin, almost at birth, to exercise their feet, by opening and closing the
toes, so as to perch and walk with them later.   This applies to Woodpeckers,
Owls, and other birds breeding in holes, as well as to perching birds proper,
the coarse chips, rotten sawdust, and the rough character of the bottoms of
nests in hollows serving the end pointed out.

The Blue Jay is pictured so many times in connection with the story of
the growth of these birds, that with a final picture of an adult bird, only a
word as to color and size seems demanded.   Adult birds are nearly a foot in
length.   The prevailing color above is of blue of varying shades, and below
is white.   The birds have conspicuous blue crests, the forehead is black.
There is a black band across the breast, reaching up on the sides of the neck
and joining on the back of the head.   The blue wings and tail have some of
their feathers barred with black and tipped with clear white.

The Blue Jay is found throughout Eastern North America, south to
Florida and Eastern Texas and north to the Fur Countries.   It is somewhat local in its distribution, and is generally resident.   It varies greatly in
its habits as regards its association with man.   At points, as in the cases
cited, this is a common bird in large towns, breeding in trees close to houses,
and apparently the least shy of birds.   Again, in other localities it seems to
avoid the vicinity of houses, and retires to the solitude of the woods.   The
eggs are pale brown or buffy olive, sprinkled all over with minute spots of a
darker shade.   They are about an inch and a tenth long and nearly nine
tenth of an inch broad.

This is the local Southern race of Blue Jays resident in Florida and the
Florida Blue Jay. Gulf Coast region of Texas.   Smaller than our Jay, the
Cyanocitta cristata florincola colors are similar but grayer and more purple or laven-
Coues. der in tone.   The white markings on both wings and tail

are also more restricted. This is an eminently sociable bird, common in all towns and breeding in the several kinds of oaks that are in such general use as shade and ornamental trees.

The Magpie is a common species in Western North America, and has been recorded from Montreal and Illinois. They are large Jay like birds, some nineteen inches long, the tail being very conspicuous. **American Magpie.** Their prevailing colors are lustrous black, with iridescent Pica pica hudsonica (Sab.). metallic shades of deep blue, purple and greenish. The white above is confined to a broad patch on each wing. White is the prevailing color below except on the breast and throat. They build very large partially covered nests, of coarse material, and lay from six to nine eggs. These are crow-like in general color and appearance. They are more than an inch and a quarter long and nine tenths of an inch broad.

This bird is the smallest of our Flycatchers, the average length being about five and one half inches. Its general colors are olive on the upper parts, greenish or brown in tone, and grayer on the head. **Least Flycatcher.** There are *two wing bars of ashy white.* The lower parts Empidonax minimus are white, tinged with gray on the breast and sides, with Baird. sometimes a faint wash of yellow on the white of the belly. The lower mandible is dusky brown.

This is a bird of Eastern North America, breeding from Pennsylvania north into Canada and wintering in Central America. The nest of rootlets, strips of bark and plant fibres, is a compact, cup shaped structure, placed generally in fork or crotch from ten to twenty feet from the ground. The eggs are pure white without markings, rather more than three fifths of an inch long and half an inch in their smaller diameter. This is a common species, frequenting orchards and the vicinity of cultivated grounds rather than the woods. It has a jerky, imperative call, which has been well described by the word "chebec," so characteristic as to have become one of the common names of the bird.

In the early spring days, before the leaves appear, just after the Bluebirds and Robins arrive, comes a rather slim bird, with long tail and large head, of a general dusky olive color above, white below washed with faint yellow on the belly, and with gray brown on the sides and throat. This is

Phœbe.
Sayornis phœbe (Lath.).

the Phœbe or Bridge Pewee, which is instantly recognized by its call note—"he names himself Phœbe"—and its constantly wagging tail, which is as characteristic as his note. The bird is about seven inches long. It is very sociable, and with the advent of man has largely abandoned former nesting sites, which were presumably under banks and on ledges in caves. Now any rafter forming a shelf, whether under a bridge, on the piazza, or in the barns or outbuildings, affords him nesting sites. In such places a bulky, shallow nest of mosses and mud, lined with hairs and grasses, is located. Four to six white eggs are laid which sometimes are dotted with a few reddish spots. About three quarters of an inch long, they are nearly three fifths of an inch in their other diameter. This is a bird of Eastern North America. Breeding from the Carolinas north to Newfoundland, it winters from its southern breeding points to Cuba and Eastern Mexico.

Say's Phœbe is a Western species which has been recorded from Illinois, Wisconsin, Iowa, and Massachusetts. It is a bird in general build and size much

Say's Phœbe.
Sayornis saya (Bonap.).

like our Bridge Pewee but rather larger. Light brownish gray above, it has a blackish tail. The lower parts are much like the upper on the throat and breast, but the belly, sides, and flanks are a warm cinnamon brown. Its nesting is very similar to that of its congener, as are its eggs.

RUBY-THROATED HUMMINGBIRD. YOUNG MALE.

This is the only hummingbird of East-

**Ruby-throated
Hummingbird.**
Trochilus colubris Linn.

ern North America. It has been so frequently described and eulogized as to have become familiar to all.

With its lustrous green back and metallic ruby throat the male is certainly magnificent, and his mate, scarcely less beautiful, is very like him, save that she lacks the gorgeous throat patch, and has a more rounded tail. Young birds are very like the female, the males having the throat streaked with dusky feathers. Adult males in the fall lack or have only traces of the brilliant throat patch.

The nest is

RUBY-THROATED HUMMINGBIRD. ADULT FEMALE.

befitting the dainty bird. Made of soft plant down closely felted, it is covered outside with lichens, and matching very closely the limb on which it is saddled, it may readily be passed by or overlooked, seeming but a knot on the branch.

These nests are generally from fifteen to twenty feet from the ground and in almost any kind of tree. Two eggs, pure white and about half an inch long, are laid. The birds range throughout Eastern North America, from Florida to Fur Countries, breeding from Florida to Labrador and wintering in Cuba, Eastern Mexico, and Central America.

This is a dusky, blackish gray bird about five and a half inches long. It may be readily recognized by the peculiar elongated shafts of the tail feathers, which are very stiff, decidedly pointed, and extend beyond **Chimney Swift.** the vanes. The eye is deeply set in the head and there is Chætura pelagica (Linn.). a prominent, overshadowing eyebrow. The nest, formerly placed in hollow trees and caves, is now universally built in chimneys. It is made of dead twigs glued together and to the interior wall of the chimney with saliva. In form it resembles a shallow bowl or basket cut in half, the chimney forming the back wall. The eggs are pure white, are from three to six in number, and measure about four fifths of an inch in length and half an inch in their other diameter.

The birds arrive from the South from the middle to the last of April in the Middle and New England States and remain until well into October. They range throughout Eastern North America to the Fur Countries and winter in Central America.

CHIMNEY SWIFT.

CHIMNEY SWIFT'S NEST.

This is the smallest of our Woodpeckers in Northeastern North America, being generally less than seven inches long. Its prevailing colors are black and white. There is a clear, broad stripe of white down the centre of the back and numerous white spots on the wing feathers, also a white stripe above and one below the eye. In the adult female the white stripes above the eye are interrupted from joining by a narrow stripe of black at the back of the head. In the adult male a bright scarlet band connects these stripes. In young birds, the whole top of the head shows scarlet feathers mixed with black. The middle tail feathers are black and the other ones white with black bars crossing them. This is a resident bird throughout the middle and northern parts of the Eastern United States and northward. It breeds in holes, generally excavated in some dead limb. The eggs are pure white, three to six in number, about three quarters of an inch long and five eighths of an inch in their other diameter.

Downy Wood-pecker.
Dryobates pubescens medianus (Swains.).

This is the form occupying our more southern regions, and its chief distinctions from its more northern congener are its smaller average size and its brownish white breast. Recent investigation shows that this was the form described by Linnæus and it is so regarded by the authorities on nomenclature. It seems hardly necessary to say that in general habits it closely resembles its more northern representative, though as I have seen it in Florida it is more of a wood than a house bird.

The Southern Downy Woodpecker is the geographical race, occupying the South Atlantic and Gulf States from South Carolina to Florida and Texas.

Southern Downy Woodpecker.
Dryobates pubescens (Linn.).

There are two distinct color phases of this bird that do not correlate with age, sex, or the season of the year, and which, though in their extremes are very different, yet intergrade completely. The one extreme of color is bright tawny red, and the other silvery gray, darker on the upper parts. These colors are on a white or grayish ground. The length of the bird varies from nine to ten inches. Small owls, they have pronounced horns or ear tufts, *and are the only owls of this size in the region under consideration that are distinguished*

Screech Owl.
Megascops asio (Linn.).

SCREECH OWL. GRAY PLUMAGE

by this feature. Their eyes are yellow, of varying shade. The birds are common throughout Eastern North America as far north as New Brunswick and south to Georgia, ranging west as far as the Great Plains. They are generally resident and are more abundant in the middle districts and in the south. They frequent the vicinity of dwellings and are oftener heard than seen. The name given them locally in the south, " Shivering Owl," well describes the peculiar tremulous cadence of certain of their call notes. They breed early in the season in hollow trees and the deserted homes of the larger Woodpeckers, laying from three to six pure white eggs about an inch and two fifths long, and a little less than an inch and a fifth in their smaller diameter.

This is a near ally of the common Screech Owl and closely resembles that bird. It is found in the same varieties of color as its prototype, but it

Florida Screech Owl.
Megascops asio floridanus
(Ridgw.).

is decidedly smaller and generally darker colored, especially on the under parts. In general habits and methods of nesting the two are identical. They are found in lower South Carolina and Georgia, chiefly near the coast, and throughout Florida and Southern Louisiana.

The Barn Owls are the birds of literature, sung in song and praised in prose, birds of the ivy mantled church tower, and of old ghost inhabited

American Barn Owl.
Strix pratincola Bonap.

castles long gone to decay. The prototype of our bird throughout England and Europe is very similar in appearance, and representatives of the genus have an almost world-wide range.

Our bird varies from seventeen to eighteen inches in length and has the upper parts bright yellowish brown with a gray mixture and spotted with flecks of black and white. The tail is of similar color with dusky bands of varying distinctness. The lower parts vary from snowy to tawny and are marked more or less abundantly with dots shading from gray to almost black. There are no ear tufts, but the facial discs are very much developed. They are light colored with a distinct border of yellowish brown. The eyes are comparatively small and the iris is so dark a brown as to form little contrast to the darker pupil. *This is the only tawny owl in this region with black eyes.*

SCREECH OWL. RED PLUMAGE.

The birds nest in old buildings, steeples, and belfries as well as in hollow trees and sometimes in holes in banks. The eggs vary in number from four to eight and even nine, are pure white, about an inch and two thirds long and rather more than an inch and a quarter in their smaller diameter.

The birds are found throughout North America, becoming rare toward the northern border of the United States and ranging but little farther north. South, they are found through Mexico. They breed from Southern New York and Connecticut south.

There are two kinds of vultures that frequent the vicinity of houses both in towns and country throughout the Southeastern United States. Both **Turkey Vulture.** have bare heads and are dark in color. The one with the *Cathartes aura (Linn.).* bare skin on the head *black*, general plumage intense *and* glossy black, and tail short and very squarely cut is the **Black Vulture.** Black Vulture. Its length is about two feet, and though *Catharista atrata (Bartr.).* shorter than its ally, it is a heavier and more compactly built bird. It does not range or breed as far north as its congener, being uncommon north of North Carolina, and breeding generally to the south of that State. The eggs are faint bluish white in color with sparse markings and washes of varying shades of brown. They are about three inches long and two inches in their other diameter, and are laid generally on the ground without any nest structure proper, but in unfrequented and thick places, under bushes or palmettoes.

The Turkey Buzzard is the Vulture with the bare skin of the head *bright red*. The general plumage is black with a decidedly brownish tinge, and the tail is longer and more rounded than in the Black Vulture. About thirty inches long, it is a slimmer and lighter bird in weight. Its range is more northern, regularly as far north as central New Jersey, beginning to breed a little to the south of that State.

Its eggs are not quite as large as those of its ally. They are rather more than two inches and four fifths long and about an inch and nine tenths broad. Similar in general color, they are more thickly and distinctly marked, and are laid sometimes in a hollow log, or in a cavity in rocks, but more frequently on the ground and without any nest material. Immature birds of both kinds of Buzzards have the head and neck sparsely covered with feathers of a dusky color and downy or fur like character.

Both birds are of great use to man, being untiring in finding and con-

TURKEY BUZZARD.

suming all kinds of dead animal matter, and acting as sanitary police without regard to the wishes of other dwellers about them.

In the gardens of many southern homes and often seen walking in the sand of road or street, is a small dove, in form very like a house pigeon, but diminutive. The bird is only about six inches and three quarters long and may readily be recognized by its size alone, for it is the only pigeon approaching this size in Eastern North America.

Ground Dove.
Columbigallina passerina
terrestris Chapm.

It is a bird with the colors to which doves have given a name. Dove color is perhaps difficult to formulate in words, but is known to all. In the old male, this color is richly tinged with a wine shade on the breast where most of the feathers have blackish centres ; those of the head and breast have their edges of a darker shade of color giving a scaled appearance. The top of the head is shaded with bluish and this color gradually shades into brownish dove color on the back. There are iridescent metallic spots on some of the feathers of the shoulders. The wings are brownish gray when closed, but when the bird takes flight, the bright cinnamon of the inner webs shows plainly. The tail is dark brown, the outer feathers having narrow white tips. The female is similar to her mate, but generally paler and more grayish in tone. The young at first are like the female but much duller, and the feathers are marginal with lighter shades of their ground color, sometimes almost white. The birds are found throughout the Southern States from North Carolina south. They are generally resident and breed throughout their range, building a characteristic pigeon nest of dead twigs loosely laid together with now and then some straws or even a few feathers added and placed on the ground or in a low bush. Two white eggs are laid rather more than four fifths of an inch long and two thirds of an inch in their other diameter.

This is a bird about ten inches long, which looks very like the Mourning Dove, except that it has a square tail, the feathers of which are tipped with ashy gray. It is found on the Florida Keys, but is not common, being much more abundant in the West Indies and the Bahamas. Its nest is on or near the ground, and two white eggs are laid rather larger than those of the Mourning Dove.

Zenaida Dove.
Zenaida zenaida (Bonap.).

The wild pigeons having almost disappeared from the Eastern United States, the representative of the family left in the northeastern portion is the Mourning Dove. So that it is pretty safe to say that here **Mourning Dove.** the only pigeon-like bird to be met with in a wild state is Zenaidura macroura (Linn.). this bird. It is quite familiar, going about in twos or threes, breeding often in evergreens or thick bushes close to our houses, and dusting itself in the roads about the localities it frequents. It is about a foot or rather more in length and is clothed in traditional dove color. Bluish on the crown, with iridescent sheen and of a general olive dove color on the back and rump, it is to be known from its size and a *small, black, crescent-shaped mark* below the side of the face. It has the wedge-shaped tail so characteristic of the Passenger Pigeon and is like him in general appearance but is much smaller.

MOURNING DOVE.

The nest is a loose structure of dead twigs, placed in a tree generally well up, some ten feet or more, and but rarely on the ground. The two white eggs are an inch and a fifth long and more than four fifths of an inch in their smaller diameter. This dove is found throughout temperate North America from Southern Canada and Maine south to the West Indies and Panama. It breeds from Cuba north and winters from southern New Jersey and Illinois south.

This is a large pigeon, fourteen inches long and dark in general color, resembling closely a certain color phase of the Domestic Pigeon, deep slate with greenish and purple sheen. The entire top of the head above the eyes is

pure white in 'the adult male, and gives the bird a characteristic on which its
name is based and by means of which it is readily recognized.   The female

**White-crowned
Pigeon.**

Columba leucocephala
Linn.

bird is like the male but is duller, and has the white of the
head ashy.

This pigeon is a bird of the Greater Antilles and is of
regular occurrence at Key West and on the Keys off the
southern end of Florida.   It nests in low bushes, lays two white eggs, nearly
an inch and a half long and an inch in their smaller diameter.

WHERE ROBINS SING AND NEST.

ALONG THE HIGHWAY.

ALONG THE HIGHWAY.

THE Red-poll Warbler is about five inches and a quarter long. The adult birds are distinguished by a deep chestnut brown crown. The back is olive brown, with a distinct grayish wash or suffusion, shading into olive green on the rump. There is a pale yellow line above the eye, defining the crown patch. Below, the throat and breast are bright yellow, shading into grayish white on the belly. The feathers below the tail are yellow. The sides of the throat, breast, sides, and flanks are streaked with bright chestnut. The tail is dusky or blackish, the outer feathers having white areas on their inner webs. In the fall and winter the birds have the crown obscured by the dull brown tips of each feather, and the chestnut color is sometimes replaced by brown. The line above the eye is grayish or whitish. The breast is dusky, and the rest of the under parts are grayish white washed, more or less, with faint yellow. The breast and sides are streaked more obscurely than in summer with dusky brown.

Palm Warbler.
Dendroica palmarum
(Gmel.).

The birds build, on or near the ground, a nest of grasses lined with finer grasses and plant fibres. Four or five eggs are laid, dull white with some cinnamon and brown markings mainly about the larger end. They are about two thirds of an inch long, and half an inch broad.

These birds breed in the interior north of the United States to the Great Slave Lake region. Their migration is chiefly through the Mississippi Valley and west of the Alleghanies, though a few pass regularly through the more northern Atlantic States. They winter in the South Atlantic and Gulf States, in the Bahamas and West Indies and in Mexico.

The Yellow Red-poll Warbler is a close ally of the bird just described. It is found in Eastern North America north to Hudson's Bay, breeding from Nova Scotia to that point. It passes through the Atlantic States, in num-

bers, in spring and fall, and winters in the South Atlantic and Gulf States.
It is not as abundant in Florida in the winter as the Red-poll Warbler.

Yellow Palm Warbler.

Dendroica palmarum hypochrysea Ridgw.

It is larger than its more western prototype, and brighter yellow prevails throughout its plumage. The adults in spring and summer have bright chestnut brown crowns. The back is olive green, with a brownish tone shading into olive green on the rump, and exposed edges of the tail feathers ; the outer tail feathers have areas of clear white on their inner webs near their ends.

A line over the eye defining the crown and a ring about each eye is bright yellow. The lower parts are the same bright yellow throughout. The sides of the throat, the breast, sides, and flanks are streaked with bright chestnut. In winter the crown patch is dulled by the brown tips of each feather, much as in the Red-poll Warbler, but the eye ring, the line over the eye, and the ground color of the entire under parts is bright yellow. All of the yellow areas are often suffused with ashy gray, and that color also obscures the chestnut markings on the yellow ground which are disposed the same as in individuals in summer. The nesting is practically the same as that of the Red-poll.

These two warblers affect open country, and are to be met with in low bushes along highways or in fields. In Florida, in the winter, the streets of the towns and villages seem their choice hunting ground, and a peculiar wagging of the tail up and down is a constant and characteristic motion. I met them in the main streets of Kingston, Jamaica, West Indies, where they were common birds.

The village street well shaded by high trees is the chosen haunt of the Warbling Vireo. In such localities, particularly where elms and maples abound, his voice may be heard high in the upper branches.

Warbling Vireo.

Vireo gilvus (Vieill.).

The birds are found throughout North America, in the eastern region from Florida to Newfoundland. They breed throughout most of their Eastern North American range, and winter in tropical America.

The Warbling Vireo looks very much like the Red-eyed Vireo, but is a much smaller bird, only about five inches and four fifths long. The same olive green shades prevail on the upper parts, but they are decidedly *lighter* and *much grayer in tone*. This grayish tone is most pronounced on the fore-

WARBLING VIREO.

head and top of the head, and gradually shades to a brighter olive on the lower back and rump. The wings and tail are dusky gray, and the exposed edges of the feathers of the tail and of the larger feathers of the wing are olive green. There is a well defined grayish white line above the eye and reaching well back on the side of the head. The lower parts are whitish, washed with faint yellowish on the breast, and more distinct greenish yellow on the sides, flanks, and feathers below the tail. The birds build pensile or semi-pensile nests of various vegetable fibres and grasses well woven together. The location chosen is a fork on an outer branch, sometimes low but generally high up in the tree. Four white eggs are laid. They are sparingly marked about the larger end with dots and flecks of dark brown almost black. The eggs are about three quarters of an inch long and upward of half an inch broad.

Lincoln's Finch or Sparrow is about five inches and three quarters long. The upper parts are brownish olive, of a rather gray tone, very definitely streaked with black. The top of the head is purer deep **Lincoln's Sparrow.** brown streaked with black and a line of gray divides the
Melospiza lincolnii (Aud.). crown in the centre. The lower parts are white streaked narrowly with black, particularly on the sides and flanks. There is a broad band of light buff across the breast, a stripe of like color on either side of the throat, and the sides and flanks are washed with a similar shade.

These birds nest in a manner similar to their congeners the Swamp and Song Sparrows, and lay four or five pale green or light buff colored eggs heavily spotted and marked with reddish brown and dark lilac. The eggs are less than four fifths of an inch long and a little less than three fifths of an inch broad.

This Sparrow is found throughout North America at large. They breed in Eastern North America from Northern New York and Northern Illinois northward. They winter from Southern Illinois south to Panama.

Lincoln's Finch is not a common bird in the States east of the Alleghany Mountains, but is of regular occurrence during the migrations in spring and fall. With many wren-like habits it avoids observation, stealing along through this or that tangle of bushes and weeds about the edges of fields and swamps and in stone walls and thickets along the highway. The birds are therefore probably more common in the Atlantic States than is generally supposed.

The Grass Finch is a stout rather heavy bird about six inches and an eighth long, of a general grayish brown color above, showing when he flies an *outer white feather* on *either side of the tail.*

Vesper Sparrow. The upper parts are brown, with a decided gray tone, *Poocætes gramineus (Gmel.).* and are streaked with black and yellowish buff.

The wings are dusky. There are two white bars on each wing and the shoulders are bright reddish brown. The outer feather on each side of the tail is pure white on most of its surface, and there is more or less white on the next feather. The rest of the tail feathers are dusky or blackish, showing grayish brown on their exposed edges. The lower parts are white, streaked with dusky brown or blackish on the sides of the throat, and on the breast, sides, and flanks.

VESPER SPARROW.

In fall the birds are darker above, the white wing bars are more or less buffy, the reddish brown of the shoulders is not so bright, and the sides and flanks are washed with pale reddish brown.

The birds build on the ground. The nest is made of coarse grasses and lined with finer grass and hair. Four or five eggs are laid. They vary in ground color from white to pale pinkish brown, and are evenly spotted and specked with reddish or dark brown. They are about four fifths of an inch long, by three fifths of an inch broad.

The Grass Finch frequents dry upland fields. You are almost sure to see him running along ahead of you in the well worn paths on the grassy

BALTIMORE ORIOLE'S NEST

sides of roads. Presently he flies to the fence showing you his characteristic tail, which is spread a little in flight. Now if it is June and late in the day you may hear one of the purest of bird songs, that has earned for its maker the title of Vesper Sparrow.

This Sparrow is found throughout Eastern North America to the Plains. They breed from Virginia and Southern Illinois north to Nova Scotia and Ontario. They winter from Virginia southward.

Passing up the village street when the stately elm is at its best in early June, you are almost sure to hear a series of bold gay whistles. Presently there is a flash of bright orange through the green foliage **Baltimore Oriole.** and with a curious resonant wren-like chatter you see the
Icterus galbula (Linn.). Baltimore Oriole alight near his inconspicuous mate. She

BALTIMORE ORIOLE. ADULT MALE.

YOUNG BALTIMORE ORIOLES.
ON LEAVING NEST. ABOUT TWO WEEKS OLD.

93

BALTIMORE ORIOLES. ADULT MALES.

would perhaps have escaped your attention but for his advent, for the nest she is building is well protected[1] by the leaves about it and its own neutral color, though it hangs at the extremity of the drooping limb above your head.

The male bird is about seven inches and a half long. When fully mature his head is covered by a black hood reaching to the breast in front and well down on the back above. The breast, sides, belly, and the lower part of the back are bright fiery orange. The wings are black with orange shoulders and a single distinct white bar. The tail feathers are orange, each one having a black area and forming together a broad wedge shaped mark, on the bright orange ground.

The female is dull brownish orange or yellow above and below. This color is relieved by some admixture of blackish feathers on the head, and sometimes by a black spot on the throat. The tail is dull brown orange, or brownish yellow, the middle feathers of which are obscurely marked with dusky or black. The wings are dusky, and some of the feathers are margined with whitish.

Young birds resemble the female, but have no black at first, and are more olive yellow in general color.

The nest is figured on a preceding page and is a marvel of industry and skill. Both birds work at weaving it but the female is the director of work and the chief laborer. These nests are usually suspended from stout twigs near the extremity of the limb or branch, and from fifteen to fifty feet from the ground. They are made of various plant fibres, stray horse hair, and fine strips of bark.

From four to six white eggs are laid. These are curiously marked with angular scrawling and a few spots or dots of dusky brown or umber. They are a little over nine tenths of an inch long, and more than three fifths of an inch in width.

The Baltimore Oriole is found in Eastern North America extending west nearly to the Rocky Mountains. It breeds from the Gulf States to New Brunswick and winters in Central America.

IN THE WOODS.

IN THE WOODS.

THIS is the Thrush famed for his vocal powers, whose song the poets have immortalized. It is smaller than the Wood Thrush, and not so robustly built. The color, beginning at the head, is a fine olive, which extends on the neck and back, shading into cinnamon brown and *bright reddish brown on the tail.* The spotting on the throat and under parts, which

Hermit Thrush.
Turdus aonalaschkæ pallasii
(Cab.).

are white with a buffy suffusion, is not so defined or extensive as in the Wood Thrush. These spots are of two kinds, being arrow shaped at the ends of the feathers, on the sides of the throat, and round in the centre of the feathers of the breast. The sides and flanks are brownish olive gray, and indistinct spots appear on some of the feathers, where the color of the sides shades into the pure white of the body. The sexes are alike. The young are more profusely and obscurely spotted below and have the feathers of the upper parts streaked with light buff or dull white. Adult birds are about seven inches long.

The nest is built of leaves and coarse grasses lined with finer material. It is placed on the ground, and three or four pale greenish blue eggs are laid. These are nearly nine tenths of an inch long and about two thirds of an inch in their smaller diameter. The bird while perhaps the most abundant of our thrushes during its migrations, the earliest to come in the spring, and remaining the latest of the group in the fall, is retiring in its habits, preferring the solitude of the deep woods, and is therefore not so well known or so often seen.

It is found throughout Eastern North America, breeds from the Northern United States northward, and southward in the higher Alleghanies into Pennsylvania. It winters from southern New Jersey and southern Illinois to the Gulf States.

94   HERMIT THRUSH.

The Olive-backed Thrush averages a little smaller than the Hermit Thrush. The upper parts are continuous olive, including wings and tail. A

Olive-backed Thrush.

Turdus ustulatus swainsonii (Cab.)

ring around the eyes is buff and the sides of the face from the bill have a ground of clear bright buff with brownish or dusky streaks. The under parts are white with a suffusion of buff, the spots on the throat are similar to those of the Hermit Thrush, being arrow shaped at the tips of the feathers. Those of the breast are round and *at the extremity* of the *feathers*. The belly is white and the sides grayish olive brown.

This bird has much the same breeding range as the Hermit but begins to breed rather farther north. It appears in the Middle States much later in the spring and earlier in the fall, and winters in the West Indies, Central and South America.

Its nest is built in low bushes or small trees, near the ground, and is made of mosses, grass, leaves, strips of bark and fine roots. The eggs are bluish green specked with cinnamon. They are nine tenths of an inch long and a little more than three fifths in their other diameter.

The Gray-cheeked Thrush is the largest of this group excepting the Wood Thrush, about seven inches and a half long. Its prevailing color is

Gray-cheeked Thrush.

Turdus aliciæ Baird.

rich olive with a grayish tone including wings and tail. The tone of the eye ring and the sides of the face is ashy gray sometimes approaching dirty white. The belly is white and the breast is spotted with half round spots at the edge of the feathers.

It is a more northern bird than either the Olive-backed or Hermit Thrush, its southern breeding point being Labrador on the eastern coast. It arrives late in the spring and passes south again in late September and early October. It winters in Central America.

Its nesting is similar to the Olive-backed in low bushes; the four greenish blue eggs are somewhat larger and spotted with reddish brown.

Bicknell's Thrush is a geographical race of the Gray-cheeked Thrush, which it resembles in general appearance. It has the same plainly colored

Bicknell's Thrush.

Turdus aliciæ bicknelli (Ridgw.).

face, but averages more intense in color. It is about the size of the Olive-backed Thrush, a little less than seven inches in length.

OLIVE-BACKED THRUSH.

This Thrush breeds in the higher elevations of the Catskill and White Mountains, and in Nova Scotia. It winters in Tropical America.

BICKNELL'S THRUSH.

The nesting economy is like that of the Gray-checked Thrush, but the eggs are smaller, greener, and the spotting is finer.

Wilson's Thrush is a bird pale golden brown on the upper parts including wings and tail. The under parts are white with a buff suffusion on sides **Wilson's Thrush.** of throat and breast, which are marked with wedge shaped Turdus fuscescens Steph. spots *a little lighter in tone and of the same color as the back.* The sides and flank are tinged with grayish. The birds are about seven inches and a quarter long.

The nest, of dead leaves, strips of bark lined with fine roots and grasses, is placed on or very near the ground. The eggs are very like those of the Wood Thrush in color and are unspotted. They are less than nine tenths of an inch in length and rather more than two thirds of an inch in their smaller diameter.

The birds are found in the Eastern United States, to the northward as far as Newfoundland and Manitoba. They breed from Northern New

98   NEST OF WILSON'S THRUSH.

Jersey north, and south to North Carolina on the Alleghanies. They winter south of the United States.

The song of the bird is very musical and reminds one of the finishing strokes of the whetstone on a good ringing scythe blade. Its call note is a clear whistle. The bird is eminently of the woods, preferring unfrequented localities, particularly the shade of dark places.

The geographical race of Wilson's Thrush that is found in the Rocky Mountain region is known as the Willow Thrush. It closely resembles its

Willow Thrush.
Turdus fuscescens
salicicola (Ridgw.).

ally, but the upper parts are *noticeably darker*, and the chest is paler buff. It is found during its migrations regularly east as far as Illinois and casually in South Carolina.

The Red-winged Thrush is a European bird that has been recorded from Greenland. It is about eight inches and a half long.

Red-winged Thrush.
Turdus iliacus Linn.

It is a plain brownish bird above with a whitish stripe over either eye. The throat, breast, and belly are whitish streaked with dusky brown. The sides and flanks are light reddish brown. The sexes are alike.

The Varied Thrush is about nine inches and a half long. Its general appearance is robin-like. The male is dark leaden brown above. This color

Varied Thrush.
Hesperocichla nævia (Gmel.).

is broken by a stripe of orange brown above the eyes and reaching back over the region of the ear. The same color is conspicuous on the shoulders, forming two bars. There is also a patch of similar shade at the base of the larger wing feathers. There is a broad band of black or dusky slate across the breast which reaches up to either eye. This defines the orange brown throat. The color back of the dark breast band is orange brown, which gradually shades into white on the belly and feathers below the tail. The female is much paler in tone than the male. The nest is a typical thrush structure of twigs, mosses, and grasses. It is placed in a low tree or bush. The eggs are pale greenish blue, sparsely marked with brown spots. They are about an inch and a tenth long and somewhat more than four fifths of an inch broad.

WILSON'S THRUSH.

In the Woods.

These birds are found in Western North America. They are more common on the Pacific Coast. They breed chiefly north of the United States, wintering as far south as California. There are records of the Varied Thrush from New Jersey, Long Island, and Massachusetts.

The Blue-gray Gnatcatcher is a small slim bird deriving much of its apparent size from its tail which is two inches long, the entire bird being but **Blue-gray** about four and a half inches in length. Its name fairly **Gnatcatcher.** indicates the general color and the prevailing tones of the Polioptila cærulea (Linn.). bird are bluish gray. There is a narrow black band on the forehead just back of the bill, which extending back defines the fore part of

BLUE-GRAY GNATCATCHER.

the crown. The middle tail feathers are black and the outer ones white, while the intermediate ones grade from the black of the middle to the white of the outer ones. The blue gray of the general plumage shades to white on the belly. The wings are dusky with blue gray edging the feathers.

This is a bird of southern distribution, rarely migrating north of New Jersey and Connecticut on the coast, but in the interior extending its journey to the Great Lakes. It breeds throughout its United States range though but sparingly toward the more northern parts indicated. It winters from Georgia and the Gulf States south. The birds build a very elegant cup shaped nest, usually saddled on a horizontal limb, but sometimes in a crotch, about thirty feet from the ground. The structure is composed of fine grasses, delicate strips of bark and plant fibre, compactly woven, and decorated externally with lichens, much after the method of the Ruby-throated Hummingbird. The four or five eggs are white much marked with spots of reddish brown. They are rather more than half an inch long, and more than two fifths of an inch in their smaller diameter.

These two diminutive birds are generally similar in appearance, and do not vary greatly in size. The Ruby-crowned Kinglet is slightly the larger,

Ruby-crowned Kinglet.
Regulus calendula (Linn.).

Golden-crowned Kinglet.
Regulus satrapa Licht.

being four and two fifths inches long. The Golden-crowned rarely exceeds four inches. Both have a coat of pale olive green, often with a grayish tone. The wings and tail are more dusky, and the feathers are edged with lighter olive green. The lower parts of both species are obscure whitish, often tinged with olive.

The Ruby-crowned Kinglet has the olive green coat extending unbroken to the tail, concealing in adult male birds the bright scarlet patch on the crown. The females and immature birds lack this bright color on the head.

RUBY-CROWNED KINGLET.

103   GOLDEN-CROWNED KINGLET.   ADULT MALE.

The Golden-crowned has a much more elaborate coloration about the head. The region about the eye in an adult male is black, then comes a stripe of white tinged with olive, extending back below a black stripe about equal in width, forming a distinct border to the crown except on the back of the head. Next to the black is a lemon yellow stripe about the same width as the other stripes, and the crown of the head inside these stripes is deep cadmium yellow. The head of the female is like this except that it lacks cadmium, the entire crown being lemon yellow.

The nest and eggs of the birds are essentially alike. A pensile or semi-pensile nest of mosses, strips of soft bark, and rootlets is suspended in a fork in some coniferous tree, at varying heights up to sixty feet. It is well felted, and woven and lined with feathers. From five to nine and even ten eggs are laid. These are creamy white in ground color, speckled and blotched with varying shades of reddish brown. The eggs of the Ruby-crowned Kinglet are less marked than those of the Golden-crowned. The eggs are a little more than half an inch long, and about two fifths of an inch broad.

The birds are gregarious in the migration and often associated. Both are songsters, but the vocal powers of the Ruby-crown exceed those of his congener.

They have a similar geographical distribution, breeding from the northern part of the United States northward. The Golden-crown also breeds on the higher elevations of the Alleghanies south into North Carolina. The Ruby-crown breeds chiefly north of the United States, but also in the elevated regions of the Rocky Mountains south into Colorado and New Mexico. The Ruby-crowned Kinglet winters from the Carolinas southward into Central America. The Golden-crown remains common as far north as Massachusetts, and winters from that point to the South Atlantic and Gulf States.

The Tufted Titmouse is a small leaden colored bird resembling a Jay that whistles loud and clear like a schoolboy calling to his playfellows scattered about the wood. He has a black forehead and a pronounced crest, which is lead color like the rest of the upper parts including wings and tail. The under parts are lighter gray becoming white on the belly, and the sides and flanks are washed with tawny brown. These birds are about six inches long. They breed in holes, generally those abandoned by the smaller woodpeckers, lining them

Tufted Titmouse.
Parus bicolor Linn.

with feathers, leaves, hair, and grasses. The eggs, from four to eight in number, are white, marked all over with spots of reddish brown. They are about seven tenths of an inch long, and more than half an inch in their other diameter.

TUFTED TITMOUSE.

The birds are found from Northern New Jersey and Southern Iowa south to the Gulf States. They are resident and breed throughout their range.

Eminently birds of the woods, they are gregarious, except in the nesting season. They hunt in small parties, often associated with Kinglets, Chickadees, and Nuthatches, and are among man's best friends, protecting our forests by their constant warfare on destructive insects.

Four kinds of Nuthatches are described by naturalists from the region treated of. They are all birds of the woodland, though not infrequently seen in the vicinity of houses. In habits and motions they resemble Woodpeckers, to the casual observer, but closer study soon shows the investigator that they possess attributes all their own. Climbing up the trunk of a tree and hammering is truly like the Woodpeckers, but in a moment the Nuthatch comes down the trunk head first, and presently proceeds to the tip of some limb. This last action is so like that of a Chickadee or Kinglet, as to dispel any idea of relationship to the carpenter of the woods.

The Nuthatches all have a note in common, differing in quality. Ank, ank, or Ank, ank, ank, perhaps' describes it best. This note is constantly heard when the birds are about, frequently betraying their presence, for they are small and rather inconspicuous.

All the kinds have sober mantles of slate or lead color, while below they vary as we shall see.

The White-breasted Nuthatch is the larger of the group, being about six inches long. The males have a glossy black cap extending well back on **White-breasted** the neck. The region about the eye and sides of the face, **Nuthatch.** breast, and belly are white. About the vent some reddish Sitta carolinensis Lath. brown feathers show amongst the white and the flanks are washed with a similar tint. The middle tail feathers are like the back in color, the next are black with white tips, and grade into almost white feathers with some black near their tips and bases, on the outer tail feathers. The wings are dusky, showing much of the color of the back on the exposed edges and surfaces of the feathers, many of which have whitish tips. The female is like the male, but the black of the cap is suffused to a greater or less extent with the bluish lead color of the back.

These are the representative White-breasted Nuthatches from Georgia north to the Southern British provinces, breeding and resident throughout that range.

The Florida White-breasted Nuthatch is essentially like its close relative **Florida White-** of the North. It is smaller with less white markings on **breasted Nuthatch.** the wing, and *the sexes have the cap much alike in color.* Sitta carolinensis atkinsi These birds are found in Florida and Southern Georgia, Scott. where they are resident and breed.

The Red-breasted Nuthatch is about four and a half inches long. The adult male bird has a glossy black cap, defined by a clear narrow white line **Red-breasted** just above the eye. The region from the front of the eye **Nuthatch.** to the bill is glossy black, becoming broader as it reaches Sitta canadensis Linn. back on the sides of the head as far as the crown patch. Below this black is an area of white extending on the throat and merging

107   WHITE-BREASTED NUTHATCH. MALE AND FEMALE.

into the bright reddish brown of the lower parts, which is continuous, including the feathers below the tail. ₁

The middle tail feathers are like the back in color, the next ones are black, and the outer ones are black with white patches near their tips. The wings are like the back in color, the quills are dusky, tipped and edged with the color of the back, bluish gray.

The females and immature birds are duller than the males, have more or less suffusion of bluish gray on the head markings, and vary on the under parts from buff to obscure buffy white.

The bird is found throughout North America, breeds from Northern New York and New England north, and south on the higher altitudes of the Alleghanies to North Carolina. It winters from its southern breeding limits to the southern boundary of the United States.

In the pine woods of our Southern States is found the fourth member of this group, the Brown-headed Nuthatch. Smaller than the Red-bellied Nut-

Brown-headed Nuthatch. hatch, it is about four and a quarter inches long. The cap on the head is grayish brown, and there is a more or

Sitta pusilla Lath. less defined *round spot of white* on the nape of the neck. The upper parts are bluish gray, including the middle tail feathers, and the under parts are whitish gray. The outer tail feathers are black, with grayish tips. This is an abundant species and except in the breeding season gregarious, going about in large bands.

All of these Nuthatches are essentially alike in their breeding and nesting habits, and their eggs do not vary greatly in appearance, except in size. They breed in deserted Woodpecker holes, and other cavities in trees and stumps, line such places with a nest structure of grasses and feathers, and lay from four to six or seven white eggs, spotted and flecked with brown of varying shades. Those of the White-breasted Nuthatch are about three quarters of an inch long, and less than three fifths in width, those of the Red-bellied Nuthatch are three fifths of an inch in length by less than half an inch wide, and the eggs of the Brown-headed Nuthatch are very little smaller.

This bird is a close ally, being a geographical race of the Brown Creeper found in Northern Europe and Great Britain. It is a small bird, about five

RED-BREASTED NUTHATCHES.

and a half inches long, of a general dark brown color above, striped with grayish white and varying shades of buff. There is much of this marking on the wings and a distinct wing bar of like shade. The rump is tawny reddish. The tail is in general color like the back, the feathers being bordered or edged with obscure buff ; it is stiff and the feathers are pointed, the centre pair is longest, from which the length is graduated to the outer ones. The under parts are silky white. The bill is long, pointed, and slightly curved.

Brown Creeper.
Certhia familiaris americana
(Bonap.).

BROWN CREEPER.

The sexes are alike. The nest is built in a crevice formed between a loosened section of bark and the body of a tree. It is made of strips of soft bark, small twigs, and moss. The eggs are white, speckled with reddish brown, more profusely at the larger end, and are three fifths of an inch long and a little less than half an inch wide. The birds breed from the Northern United States north, and in elevated regions farther south. They migrate in winter as far south as the Gulf States.

In general appearance these birds are wren-like, and in climbing habits resemble Woodpeckers. They are quiet, inconspicuous, and indefatigable workers in the pursuit of such insects and larvæ as frequent the bark of trees.

BROWN CREEPER ON TREE TRUNK.

The Winter Wren is a short, stout brown bird with a short tail and very distinct black barring on the flanks, lower breast, and belly. There is a line of obscure buff above the eye, and the wings and tail are barred with black. There are obscure broken bars of black on the dark reddish brown back. The throat and breast are lighter brown than the back and are obscurely barred with a darker shade of the same color. The sexes are alike.

Winter Wren.
Troglodytes hiemalis Vieill.

It is the smallest of the wrens that are found in our woodlands, being little more than four inches long.

The nest is built of twigs, moss, and grasses placed in a brush heap in the roots of a fallen tree, or like locality. The eggs, from four to seven in number, are white with a sparse speckling of reddish brown. They are seven tenths of an inch long, and half an inch in breadth.

This bird ranges throughout Eastern North America, breeding north from the Northern United States, and at elevations on the Alleghanies south to North Carolina. They winter from their southern breeding limits southward to Florida.

The Winter Wren is famed for its vocal powers, and is one of the sweetest songsters of the deep woods, during the breeding season.

The Redstart is a very sprightly bird, with the entire upper parts, as well as throat and breast, lustrous black. The terminal parts of the wing feathers, the two middle tail feathers, and the terminal third of the remainder of the tail feathers are also black. The parts of these feathers not black are bright salmon color.

American Redstart.
Setophaga ruticilla (Linn.).

The sides of the breast, region under the wings, and flanks are intense orange salmon color. The belly is white, suffused with a salmon tinge. The bill looks somewhat like that of a flycatcher and has noticeable bristles, at its base. The female and immature birds have a general resemblance to the adult male just described, the pattern of coloration being the same. The salmon regions are replaced by greenish yellow, the black areas of the upper parts are olive green with grayish suffusion, especially on the head, and the under parts grayish white, except where the greenish yellow prevails. The birds are about five and one third inches long, and rather slim in build.

The nest is a compact structure of plant fibres and lined with finer material of a like nature. It is placed usually in a crotch, but sometimes is saddled on a limb, at varying heights, from five to thirty feet from the ground. The

WINTER WRENS.

three to five bluish white eggs are spotted, usually at the larger end, with reddish and olive brown markings. They are little more than three fifths of an inch long, and a little less than half an inch wide.

REDSTART. ADULT MALE.

The Redstarts' general range in summer is throughout North America. In the area under consideration they breed from North Carolina and Kansas to the Hudson's Bay country. They winter in the West Indies and Tropical America.

This is eminently a wood bird, flycatcher-like in some of its attributes. The adult male, with its conspicuous bill bristles, is among the most beautiful of the group to which it belongs. The upper parts are **Canadian Warbler.** dark slate gray, the wings and tail more olive. The Wilsonia canadensis (Linn.). feathers of the top of the head are black with edgings of the color of the back. There is a line from the bill to the back of and above the eye of bright greenish yellow. The region in front of, below, and back of the eye is black, which color extends in a line each side of the throat to the breast, where a series of spots joining make continuous black lines, eight or ten in number, striping the breast. The entire under parts, save for this decoration, are bright greenish yellow, fading into white on the feathers under the tail. The female and immature birds are similar in pattern, but are duller, and the markings are obscure and undefined. The birds are about five and a half inches long.

The nest is generally placed on the ground in a mossy spot, or where roots afford it some shelter. It is made of an outer layer of dead leaves, strips of bark and moss, lined with fine roots. The eggs are white speckled mainly at the larger end with reddish brown. They vary from three to five in number, are about seven tenths of an inch long, and half an inch wide.

This is a bird of Eastern North America in summer, ranging north to Newfoundland and Lake Winnipeg, and wintering in Central and Northern South America. It breeds on the higher Alleghanies, as far south as North Carolina, and is generally dispersed in the breeding season from Northern Massachusetts and Michigan to its northern limits of travel.

Wilson's Warbler.
Wilsonia pusilla (Wils.).

This is another fly-catching warbler frequenting the damp deep woods near water and is a true insect hunter, taking persistent excursions on the wing, for its prey. A smaller bird than the Canada Warbler, it is but five inches long. The upper parts, in the adult male, are bright olive green, except for a black cap on the head, and the brighter olive on the forehead. Below it is bright

WILSON'S WARBLER. ADULT MALE.

greenish yellow. There are conspicuous bristles at the base of the bill. There are no defined markings, save the black crown patch. The females generally lack this mark, but are otherwise like the males. The immature birds lack this crown patch, but are otherwise like most of the female birds.

The nest is placed on the ground in swampy woods, and is made of mosses, leaves, and grasses, lined with finer material of a like nature. The eggs are creamy white, speckled with reddish brown and lavender. Four or five are laid and they are a little smaller in size than those of a Canada Warbler.

This is a bird of Eastern and Northern North America in summer, breeding from the northern border of the United States northward, and wintering in Central America.

The Hooded Warbler is so fine in appearance as to attract attention at once. Like his two congeners, bright yellow prevails in his livery, and is emphasized by a clear jet black hood over the head, neck,

Hooded Warbler.
Wilsonia mitrata (Gmel.).

and breast, broken by a band of bright yellow, reaching from the region of one ear, across the eyes and the forehead, to the back of the other ear. The rest of the upper parts, including the wings and tail, are olive green. The inner webs of the outer tail feathers are bright yellow. From the hood back, the lower parts are bright lemon yellow. The females are similar, but have the colors less brilliant and the black of the hood is often restricted on the breast and throat, and not as clearly defined. In the young females the hood is wanting, being replaced by olive green on the head and neck, and lemon yellow on breast and throat. The young males have the black of the hood more or less obscured by the narrow yellow edging of each black feather. The birds have bristles at the base of the bill. These warblers are rather more than five and a half inches long.

The nest is placed in a bush or small tree, generally in a fork or crotch, and rarely more than a few feet from the ground. It is made of strips of bark and fine roots, lined with finer roots and grasses. Four white eggs, spotted sparingly with reddish brown, are laid. They are seven tenths of an inch long, and a little more than half an inch wide.

Hooded Warblers are found in Eastern North America, north as far as Southern Michigan and Ontario in the interior, and to Southeastern New York and Connecticut on the sea-board. They breed from the Gulf of Mexico north to the points indicated, and winter in Central America.

The Mourning Warbler is a dark olive green bird, about five and a half inches long with a hood of bluish gray on the head, extending well down on the breast where it shades into black. The belly is greenish yellow. A bird of thickets and undergrowth, it is rarely seen far from the ground. The females

Mourning Warbler.

Geothlypis philadelphia (Wils.).

and young birds lack the hood, the upper parts are dark olive green throughout, and the throat is whitish shading into gray on the breast, this color shading into the yellow of the belly.

The nest is placed on the ground or near it, is made of strips of bark and other vegetable fibre, lined with fine grasses and hairs. The four white eggs are marked with reddish brown dots at the larger end, and are about seven tenths of an inch long, and rather more than half an inch in their other diameter.

This is a bird of Eastern North America, breeding throughout the mountainous parts of Pennsylvania, New York, and New England, and at lower elevations from Northern New England and Michigan northward. Its winter home is in Central America and Northern South America.

The Connecticut Warbler is a bird of general deep olive green tint above, and is rather longer than the Mourning Warbler, and of heavier build. It has

Connecticut War-bler.

Geothlypis agilis (Wils.).

a hood of bluish gray, reaching well down on the breast, becoming *lighter on the throat*. There is a decided light gray or white eye ring. The belly is yellow shading into olive green on the sides. In fall the gray of the upper parts of the hood is suffused with olive green. The female and immature birds lack the hood, and are deep olive green above, with throat and breast light grayish brown, and the belly light yellow.

The nest is built on the ground, of vegetable fibre in the carpet of soft moss that covers the earth in the region indicated below. The eggs, four in number, are white with sparse lilac brown and black spots at the larger end. They are about a trifle over three quarters of an inch long and more than half an inch broad.

In the fall, when they are often abundant, these Warblers are silent, and have certain thrush like attributes. They feed then, low down or on the ground, in shady dark places, and when disturbed fly to some limb near by, where they sit absolutely still for a few moments, before returning to feed.

They are only known to breed in Manitoba, are of uncommon occurrence during the spring migration in the Eastern United States, but are locally often very abundant during the fall migration. They winter in Northern South America.

The Kentucky Warbler is bright olive green above, except on the head where there is a black crown, the feathers of which are tipped with bluish gray. There is a yellow line from the bill to the region **Kentucky Warbler.** above and back of the eye. Below this is an area of Geothlypis formosa (Wils.). black, of an irregular triangular shape, extending down on the sides of the throat. The entire under parts are very bright greenish yellow. The female is similar, but generally duller, and the black is less clearly defined, and often has a grayish tinge. The birds are rather stout and heavy, and are about five and a half inches long.

The nest, often on the ground and always near it, is made of a layer of dry leaves, thin fine twigs and roots, and lined with finer roots and vegetable fibre. The eggs are white, speckled or spotted evenly with reddish brown, and are nearly three quarters of an inch long and less than three fifths of an inch in their other diameter. The Kentucky Warbler is found in the Eastern United States, as far north as Southeastern New York, Connecticut, and Iowa, breeding from the Gulf States north throughout its range. It winters in Central America.

This is a bird of the damp, dense woods, living near the ground, and noticeable both on account of its bright markings as well as its loud and pleasant song.

This Warbler is one of the commonest breeding birds from Kansas and Virginia north to Manitoba and Labrador. Southward on the higher Alle- **Oven-bird.** ghanies they breed to South Carolina. They winter from Seiurus aurocapillus (Linn.). Florida through the West Indies and Central America. The Golden-crowned Thrush, or Oven-bird, is about six and a quarter inches long, is olive green above, with a decided brownish tone. There is a crown spot of golden brown defined by a black line on each side of it. The under parts are white and there is a narrow black line on either side of the throat. The breast and sides are streaked with black. The sexes do not differ in appearance.

The birds breed on the ground, building a bulky, covered nest generally in the open woods where dry leaves cover the ground. The outside is dry leaves, bark, grasses, and vegetable fibre. This is lined with finer grasses and roots ; the entrance is at one side. While this structure is more frequently in an open place, its environment conceals it so effectually that it would but seldom be found save for the presence of the birds and their evident solicitude.

The four or five white eggs are marked and speckled with reddish brown of varying shades and pattern. They are four fifths of an inch long, and three fifths in their other diameter.

OVEN BIRD. GOLDEN-CROWNED THRUSH. ADULT MALE.

This is a bird of the pine woods. Rarely found in other localities. Where such forests are extensive it is often common and frequently abundant.

Pine Warbler.
Dendroica vigorsii (Aud.).

These birds are about five and a half inches long. The male bird is olive green above, suffused more or less with grayish, particularly in the autumn. The wings are dusky gray with generally lighter edgings of gray, and tinged with olive green. There are two whitish wing bars, and the inner vanes of the two outer tail feathers are white on their terminal half or third. The throat and sides are bright gamboge yellow, which grades into white on the belly and under the tail. There are dark stripes on the sides of the breast and below the eye, which sometimes extend to the flanks and chest.

The females are similar to the males. Some individuals that I have dissected are identical in appearance but they are generally duller and much more brownish gray in tone.

The birds are resident in the great pine forests of Florida, Georgia, and the Southern States, and migrate as far north as Manitoba and Maine, breeding throughout the region indicated in suitable localities. They winter from the Carolinas and Illinois southward.

The nest is built in pines, cedars, or other evergreens, usually more than twenty-five feet above the ground. It is made of bark and other vegetable fibre. The eggs are about seven tenths of an inch long and rather more than half an inch in width. They vary from three to five in number, are white with reddish brown and umber dots and markings, which are frequently grouped in a wreath or circle about the larger end.

This is a very rare bird in the United States, a dozen or more individuals having been secured during the periods of migration. These have been re-corded from South Carolina, Virginia, Missouri, Ohio, **Kirtland's Warbler.** Michigan, Wisconsin, Illinois, and Minnesota. Its sum-mer habitat and breeding economy are unknown. It has *Dendroica kirtlandi Baird.* been observed as not uncommon in the Bahama Islands in winter. It is a large Warbler, about six inches long. The upper parts are bluish gray, the back streaked with black. The region in front of the eye and a narrow band on the forehead are black. The under parts are pale yellow, becoming almost white on the throat and feathers below the tail. The sides are streaked with black. The female resembles the male but is duller throughout.

This is a bird resembling in a general way the Black-throated Green **Townsend's** Warbler, but with more black on and about the head. **Warbler.** It is a resident of Western North America and has *Dendroica townsendi* been recorded once as of accidental occurrence in *(Towns.).* Pennsylvania.

The Black-throated Green Warbler is one of the commoner migrants throughout Eastern North America. It breeds from Connecticut north **Black-throated** to Hudson's Bay, and on the more elevated parts of the **Green Warbler.** Alleghany range, south to South Carolina. It winters in *Dendroica virens (Gmel.).* the West Indies and Tropical America.

The male is rather more than five inches long, is clear olive green above, marked in some individuals with black spots on the back. The *cheeks* and a line *over the eye* are *bright yellow*, the throat and breast black. The belly is white, often with a suffusion of pale yellow, and the sides are streaked with black. There are two white wing bars, and there is much white on the inner

BLACK-THROATED GREEN WARBLER.

web of the two or three outer tail feathers. The webs of the *outer tail feathers are white to their bases*.

The females and old males in the fall have much yellow mixed with the black of the throat, and immature birds have a preponderance of yellow, the black being much obscured, or at times wanting.

They breed generally in evergreen trees, ten to fifty feet from the ground, where a nest of fine twigs, moss, and roots, lined with finer material, is built. Three to five white eggs are laid, spotted chiefly at the larger end with varying shades of brown. They are nearly seven tenths of an inch long and not quite half an inch in their smaller diameter.

The Blackburnian Warbler is about five and a quarter inches long.

Its distinctive characteristic mark, serving to identify the adult male, is
Blackburnian Warbler.
Dendroica blackburniæ (Gmel.).
the brilliant cadmium yellow throat and breast. The head and back are black. The back is streaked with white. There is a line in the centre of the black of the top of the head, one over the eye, and an area behind the ear, of bright cadmium yellow, like that of the throat and breast. The belly is white with more or less suffusion of orange. Some of the feathers of the shoulders are white, forming a patch in strong contrast to the black of the rest of the wing, and there is a preponderance of white on the inner web of most of the tail feathers. The females and immature birds are duller than the adult male, the black becoming grayish olive, and the cadmium yellow salmon buff. There is generally less white on the wings and tail, and the under parts are whiter with a yellowish tinge.

The nesting habits are similar to those of the Black-throated Green Warbler, and the four whitish eggs are a little larger than those of that bird, and more profusely spotted with reddish and olive brown markings.

The bird is found in Eastern North America, west to Kansas. It breeds at elevations in the Alleghanies, south as far as South Carolina, and regularly from Maine and Minnesota north to Labrador. It winters in Tropical America.

The Bay-breasted Warbler is rather large, being about five and three quarter inches long. The adult male has a chestnut crown bordered on the front and sides with black, and the entire throat, breast, and sides are chestnut, varying in shade in individuals. These markings will serve to identify the male in the spring. The back is ashy white with a brownish suffusion,

Bay-breasted Warbler.
Dendroica castanea (Wils.).

streaked with dusky brown or black. There are two white wing bars and a patch of white on the inner web of the outer tail feathers near their ends. The belly is white suffused with buff. The female has an olive green crown streaked with black and frequently showing traces of chestnut. The lower parts are buffy white, with some suggestion of chestnut on the throat and sides.

In the fall both sexes and the immature birds resemble the Black-poll Warblers at that season, but may be always discriminated from that species by the lighter green upper parts, and the buffy suffusion of the lower parts, which are yellowish green in the Black-poll Warbler. This buffy tinge is strongest on the flanks, and often grades into a very decided area of chestnut in these regions.

The nest is built of grasses and vegetable fibre, and is lined with hairs and plant down. The eggs are about seven tenths of an inch long, and a little over half an inch in their other diameter. They are white, with fine markings of olive and reddish brown, mainly on the larger end.

These birds are found in Eastern North America, north to Hudson's Bay. They breed from Northern New England north, and winter in Central America.

This is the only Warbler, of the region under consideration, whose prevailing color above *is bright blue*, which feature renders recognition of the male birds an easy problem. In addition the head and

Cerulean Warbler.
Dendroica rara (Wils.).

back are streaked with black, there are two white wing bars, and all but the central tail feathers have white patches on their inner webs. The under parts are white, and the sides and flanks are streaked with blue black, which color also forms a more or less defined band across the breast. The birds are four inches and a half long. The female has the upper parts suffused with olive greenish over black, and the under parts white, suffused with pale greenish yellow. Immature birds are like the female, but the olive green and yellowish suffusion are intensified.

The nest is placed in trees, higher generally than twenty-five feet from the ground. Built of fine grasses and plant fibres, secured with spiders' webs, and decorated with lichens, it is lined with finer plant fibres. The eggs are usually four in number, and are white in color with profuse markings of reddish brown. They are nearly seven tenths of an inch long and more than half an inch broad.

The birds are found in the Eastern United States, but are rare east of Central New York and the Alleghany Mountains. They range as far north as Southern Ontario, and west to the Plains. They breed from West Virginia, Tennessee, and Kansas north to Southern Ontario and Minnesota, and winter in Central America.

In the spring migration this is one of the most strikingly variegated of the Warblers, arresting the attention at once by its conspicuous coloring. The crown is bluish slate, and has a narrow white border **Magnolia Warbler.** on each side. The forehead is black, which color extends Dendroica maculosa (Gmel.). to and below and back of the eyes on the side of the face, gradually becoming broader till this color covers most of the side of the head. The upper back is black, changing into olive green with black markings showing through. The lower parts, throat, and rump are bright yellow, the breast a n d sides c o n-spicuously striped with black. The tail is black. T h e t w o central fea-thers are un-marked, the others have broad white markings on their inner webs, leav-ing the ter-minal third

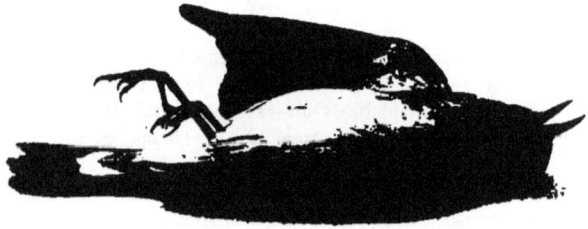

BLACK AND YELLOW WARBLER. IMMATURE BIRD IN AUTUMN.

and bases of these feathers clear black. The feathers below the tail are white, and there is a white patch on each wing. The female is at the same season duller, with its markings less sharply defined. Individuals taken in the fall are slaty gray on the head, olive green on the back, yellow on the rump,

BLACK AND YELLOW OR MAGNOLIA WARBLER.

and the black feathers above the tail are obscured by greenish olive. The lower parts are yellow with an obscure ashy belt on the chest. The belly, the feathers below the tail, and the wing bars are white. The sides show obscure black streaks. The *tail* is *like that of spring* birds, and will serve to identify these birds at any season.

They nest in much the same kind of localities as the Blackburnian and Black-throated Green Warblers, generally in evergreen trees. The nest is built of similar material but is located much nearer the ground. The eggs are four in number, white with a circle of reddish and olive brown markings about the larger end. They are a little less than two thirds of an inch long, and about half an inch in their other diameter.

The Magnolia Warbler is found throughout Eastern North America. It breeds from Northern New England and Michigan north to Hudson's Bay and south on the higher Alleghanies to Virginia. I found them breeding at Mountain Lake, Giles County, Virginia, in the summer of 1889. The altitude was over 4000 feet. They winter in the West Indies and Central America.

Black-throated Blue Warbler.

Dendroica cærulescens (Gmel.).

The Black-throated Blue Warbler is about five inches long. The males are dark grayish blue above, individuals frequently showing blackish markings on the back. The sides of the head, to and back of the eye, the throat and sides of the body and flanks are clear black. The bases of the *larger wing feathers* are white, and the inner webs of the outer tail feathers have white of varying extent near their tips. The females are olive green above, with tail and wings more dusky, and often with a slightly blue tinge. The region about the eye is dusky gray. The lower parts are obscure olive buff, grayer on the flanks and sides. The area of *white* at the *bases* of the larger wing feathers is much less than in the male, and is sometimes covered by the feathers of the shoulder. In the fall the birds are similar, the blue of the male bird is, however, tinged with olive and the black of the throat more or less obscured by the white tips of some of the feathers. The females are generally yellower than in the spring, but at all seasons and in both sexes the white bases of the larger wing feathers will serve to identify the birds.

The nest is usually built within a few feet of the ground, in the undergrowth of forests. It is composed of bark, fine grasses, pine needles, and lined with finer plant fibres. The eggs, three to five in number, are about seven tenths of an inch long, and half an inch in the smaller diameter. They

BLACK-THROATED BLUE WARBLER.

are obscure white in color, with markings of varying shades of brown, mainly at the larger end.

These Warblers are found throughout North America. They breed from Northern New England northward to Labrador. They winter in the West Indies and Central America.

Cairns's Warbler.
Dendroica cærulescens cairnsi Coues.
Cairns's Warbler is somewhat smaller than its prototype, and has the area between the wings more definitely marked with black than the Black-throated Blue Warbler. It is the local race of that bird breeding in the higher parts of the Alleghany Mountains from Virginia to Georgia.

I found this bird breeding rather commonly at Mountain Lake, Giles County, Virginia, at an altitude of about 4000 feet. This was in June, 1889. The breeding habits do not differ from those of the more northern bird.

Cape May Warbler.
Dendroica tigrina (Gmel.).
In richness of livery, few of the Warblers can compete with the Cape May. It has a kind of beauty all its own. Yellows and warm browns relieved by black and olive are combined in a pattern that defies imitation.

The birds are about five inches long. Males in the spring have a black crown, the feathers of which are just tipped with olive green. The area about the ear is reddish chestnut, and is enclosed in a larger bright yellow patch on the sides of the neck. The upper back is dark olive green, streaked with black. The rump is bright greenish yellow, and the under parts, except the white of the belly and feathers under the tail, are bright gamboge yellow streaked with black. The throat is sometimes tinged with reddish chestnut like the region of the ears. There is a large white area on the wings, and on the inner web of the exterior tail feathers. The bill is more acute than in other representatives of this genus.

The female is grayish olive green above with a yellowish rump, and a yellow stripe over the eye. Below the ground color varies from whitish yellow to sometimes deep yellow streaked obscurely with dusky. In autumn the markings of both sexes are much suffused and obscured by grayish olive.

The nest is semi-pensile, of grasses and fine twigs and roots, fastened with spider webs and fine plant fibres. It is attached to a low branch in open woods, and sometimes in isolated trees in fields.

In the Woods.

The eggs are white, dull in character, and often buffy. They are sparsely speckled, mainly at the larger end with brown spots of varying shades. They are seven tenths of an inch long and half an inch in width.

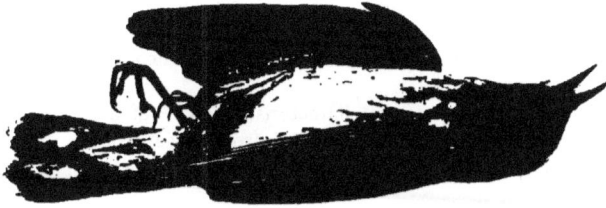

CAPE MAY WARBLER. FEMALE IN AUTUMN.

These birds are found throughout Eastern North America, north to Winnipeg and Hudson's Bay. They breed from Northern New England north throughout their range, and winter in the West Indies and Central America.

This small and agile Warbler, about four inches and three quarters long, is a common bird during its migrations, but is local in its distribution during the nesting season. It is generally found throughout North America as far north as Canada, breeds locally throughout New England, New York, and westward in the northern border States, and northward as indicated. It winters from Florida through the West Indies southward.

Northern Parula Warbler.
Compsothlypis americana usneæ Brewster.

The general color of the upper parts is bluish gray, being more blue on the head *and black in front of the eye*. The back is *bright greenish yellow* forming a *definite patch*. The throat and breast are yellow, the throat is tinged more or less deeply with orange brown in the form of a spot or band. The chest is spotted with rich chestnut. The belly and feathers below the tail are white. The sides and flanks are brownish, washed with gray. There are two white wing bars, and the outer tail feathers have a white patch near their ends. The female is similar to the male but paler, and immature birds have the upper parts suffused with greenish yellow, the yellow of the lower parts is paler and the throat often grayish white.

The nest is built of tufts of the moss and lichens that hang from the limbs of trees, or in bunches of dead leaves and driftwood left in lower branches by some passing freshet. The eggs are white, speckled particularly at the larger end with reddish brown. They vary, from three to five in number, and are rather more than three fifths of an inch long, and less than half an inch wide.

The Parula has many of the characteristic actions of a Titmouse, and is especially active, now hanging head down from the end of some twig, now fluttering at its extremity, busy always in pursuit of its insect prey.

Parula Warbler.
Compsothlypis americana
(Linn.).

This is the form of Parula Warbler breeding in the region from the District of Columbia southward, on the coast, and in the interior from Southern Illinois southward. In the Gulf States it ranges and breeds east of Texas. It is a little smaller than its northern prototype. The adult male has more yellow below and less black or dusky in front of the eye. The orange brown color band on the breast is narrower, and in many individuals obscure or wanting. The chest is light brown without distinct markings.

In habits and breeding economy the birds are essentially similar.

Tennessee Warbler.
Helminthophila peregrina
(Wils.).

The Warblers that we have so far met with are birds that have bills rather like those of thrushes. The group, of which the Tennessee Warbler is fairly representative, have acute conical bills, more like those of the orioles. The Warblers with such bills naturally form themselves into two groups, one olive and the other parti-colored. Three birds, the Tennessee Warbler, the Orange-crowned Warbler, and the Nashville Warbler, form the olive group. The parti-colored group contains the Golden-winged, Lawrence's, Brewster's, the Blue-winged Yellow, and Bachman's Warblers.

The Tennessee Warbler is about five inches long and is bright olive green on the back and rump. The head and sides of the face are in sharp contrast to the rest of the upper parts, being ash gray with a bluish tinge, which becomes grayer on the sides of the face, and there is a stripe over the eye of grayish white. The wings and tail are dusky, with olive green edgings to the feathers. The under parts are white, more or less suffused with yellow, the sides and flanks are grayish, the belly and feathers under the tail are white. The female is like the male, but has the gray of the head suffused with greenish olive, and is yellower beneath. Immature birds are bright olive green above, including the top and sides of the head. Their under parts are much suffused with yellowish, except the feathers below the tail, which are white.

TENNESSEE WARBLER.

The nest is built of fine plant fibres and moss, lined with finer material and hair, and is usually placed in bushes near the ground. The eggs, generally four in number, are white, with a circle of brown dots of varying shades at the larger end. They are about three fifths of an inch long, and half an inch in their other diameter.

The birds are found throughout Eastern North America, during their migrations, and breed from Northern New York and Maine northward to Hudson's Bay. They winter in Central America.

This Warbler is about the size of the Tennessee Warbler, is olive green with a suggestion of grayish suffusion on the upper parts. Its distinctive marks are a more or less *concealed crown patch* of *brownish orange*, and a yellow eye ring. The lower parts are greenish yellow, grayer on the sides and flanks, and with an obscure striping of dusky olive, most marked on the breast. The feathers below the tail are yellow. The sexes are alike. The immature birds lack the crown patch, are more suffused with grayish, both above and below, and have the same obscure breast stripes and yellow coloring on the feathers beneath the tail.

Orange-crowned Warbler.

Helminthophila celata (Say).

The nest is of leaves and plant fibres, built on the ground or near it, and the eggs, about the same size as those of the Tennessee Warbler, are white, speckled especially at the larger end with reddish brown.

The birds are found during their migration throughout Eastern North America, but are rare east of the Alleghanies north of Virginia. They breed in British Columbia and northward, and winter in Florida, the other Gulf States, and Mexico.

The Nashville Warbler is smaller than either the Orange-crowned or Tennessee Warbler, being about four inches and three quarters in length. Its back and rump are similar in color, bright olive green, and the top and sides of the head are grayish blue with a more or less *concealed crown patch* of *deep chestnut*. The lower parts are pure gamboge yellow, becoming lighter or almost white on the belly. The feathers below the tail are yellow like the throat and breast. The sexes are similar, though the lower parts are often not as bright in the females. Immature birds lack the bluish gray of the head and the crown patch, being olive green with a brownish suffusion above. The sides of the head are grayish, and there is a white ring around the eye. The yellow of the under parts is brightest on the throat and breast, the sides are much suffused with brownish, and the feathers under the tail are yellow.

Nashville Warbler.

Helminthophila rubricapilla (Wils.).

The nest is built on the ground in sparse woods. It is made of plant fibres and mosses, and lined with fine grasses, plant fibres, and roots. The four or five eggs are white with profuse speckling of reddish brown, mainly about the larger end. They are nearly the same size as those of the Orange-crowned Warbler.

The birds are found during their migrations throughout Eastern North

America. They breed from Connecticut and Illinois north to the Fur
Countries, and winter in Central America.

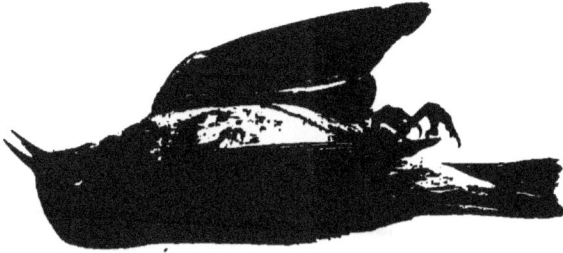

NASHVILLE WARBLER.

The parti-colored group of these sharp-billed Warblers includes the
Golden-winged, Lawrence's, Brewster's, Blue-winged Yellow, and Bachman's
Warblers.

Golden-winged Warbler.
Helminthophila chrysoptera (Linn.).
The Golden-winged Warbler is found during the
migrations throughout the Eastern United States, north
as far as Minnesota and Michigan, Ontario and Vermont.
It breeds from Northern New Jersey and Indiana northward, and south on
the Alleghanies to South Carolina, wintering in Central and Northern South
America. It is a bluish gray bird above, with a *golden yellow* crown. The
side of the face about the eye is black, separated from the conspicuous *black
throat* by a narrow white line. There is a white line above the eye, and two
bars on the bluish gray wings overlapping form an area of golden yellow like
the crown. The breast, belly, and feathers below the tail are white, and the
sides and flanks are grayish. This general pattern obtains in the female, but
the black of the face and throat becomes gray, and the yellow where it occurs
is duller.

The nest is built on the ground, of an outer layer of large dead leaves
and bark, lined with finer material. The location generally chosen is the edge

of a clearing or in a new second growth of woods. The eggs, varying in number from four to six, are white, speckled mainly at the larger end with reddish brown. They are rather more than three fifths of an inch long and half an inch in their other diameter.

The Blue-winged Yellow Warbler is well described by its name. It is bright yellow below except the feathers below the tail which are white. The crown is bright yellow, gradually shading into the yellowish green of the upper parts. There is a conspicuous *narrow bar of black, reaching from the bill to the region back of the eye.* The wings and tail are dark grayish blue, lighter on their edges. There are two bars on the wings which vary from pure white, the general color, to bright yellow. The three outer tail feathers are noticeably marked with white on their inner webs. The adult female and immature birds are similar, but have the yellow crown more or less obscured by yellowish green like the back. The birds are about four inches and three quarters long.

Blue-winged Warbler.
Helminthophila pinus
(Linn.).

BLUE-WINGED YELLOW WARBLER.

During the migration they are found throughout the Eastern United States, north to Southern New England and Southern Minnesota. They breed throughout the northern half of this area and winter in Central America.

BLUE-WINGED YELLOW WARBLER'S NEST AND EGGS.

They nest on the ground in a manner similar to the Golden-winged Warbler, and the eggs are very much like those of that bird in size and general appearance.

There are two forms of Warblers that present characters common to both of those just dealt with, the Blue-winged Yellow and the Golden-winged Warblers, that seem to demand further study before their status in the bird world can be definitely determined. These birds are known at present as Brewster's Warbler, Helminthophila leucobronchialis, and Lawrence's Warbler, Helminthophila lawrencei.

Brewster's Warbler.
Helminthophila leucobronchialis (Brewst.).

Brewster's Warbler is a bird of which over one hundred individuals have been captured and of which representatives may be seen in almost any large collection of birds. The upper parts, including wings and tail, are bluish gray, it has a more or less defined yellow crown, and a black bar through the eye like the Blue-winged Yellow Warbler. The *wing bars are yellow*. The lower parts are *white*, with a faint yellowish tinge on the breast. The tail feathers have markings of white like the Blue-winged Yellow. The sexes are similar, the females *having lighter yellow or white wing bars*.

There are many gradations of tint and color in the birds known and the divergence seems to be from the extreme type known as Brewster's Warbler, to birds more closely resembling the Blue-winged Yellow Warbler. These birds have been taken throughout the range of the Blue-winged Yellow Warbler but they are perhaps more common in Northern New Jersey near the Hudson River, in Westchester County, New York, and in the lower Connecticut Valley.

Lawrence's Warbler.
Helminthophila lawrencei (Herrick).

The author's personal experience with Lawrence's Warbler is limited to a single female bird secured by him in the region back of Tarrytown, New York, and catalogued as follows: 11,421. "Helminthophila lawrencei, female adult, Pocantico Hills, Westchester County, New York, 19 May, 1891." He has also personally examined the type bird from which the form was described, and which is in the collection of David B. Dickinson, Esq., of Chatham, New Jersey. This bird is much rarer than Brewster's Warbler, as only some half dozen specimens have been recorded.

The male looks like the Blue-winged Yellow Warbler, but has a black patch on the side of the face and a black throat like the Golden-winged Warbler. The female referred to had obscure markings of dusky, taking the place of the throat and cheek markings, otherwise the bird was like a female Blue-winged Yellow Warbler.

Bachman's Warbler.
Helminthophila bachmanii (Aud.).

This is a common migrant in the Southeastern United States. It has been secured in the vicinity of New Orleans, and at Charleston, South Carolina, but is best known as a migrant in Northwestern Florida in March and April, and at Key West in July and August. But little is known as to its breeding habits, or of the exact locality where the birds nest and rear their young. A single nest was described and attributed to this species from St. Simon's Island, Georgia, 30 April, 1854.

Since writing the above the summer home and breeding habits of Bachman's Warbler have been discovered and described. The birds are common summer residents in the St. Francis River region of Southeastern Missouri, and Northeastern Arkansas. The nest was found in blackberry bushes, two feet from the ground. It is " composed externally of dried weeds and grass stalks ; internally of fine weeds and grass stalks, lined with black fibres, apparently dead threads of the black pendant lichen," common to the region. The eggs were three in number, immaculate white, rather more than three fifths of an inch long and a trifle less than half an inch wide.

The birds are about four inches and a quarter long. The male has a yellow forehead, and the lower parts are yellow, except a black patch on the breast extending to the lower part of the throat. There is a black area on the fore part of the crown, back of which follows an area of dull blue gray. The back and rump are bright olive green. The tail is dusky gray, the two middle feathers are unmarked, and the other feathers have white patches on their inner webs near the tips. There is an area of yellow on the shoulders. The female is similar to the male above, but lacks the black on the crown, which is grayish, and the yellow on the forehead is often obscure. The under parts are white washed with yellowish on the throat and breast.

The prevailing colors of the Worm-eating Warbler are dull greenish olive above, except on the head ; the tail and wings are darker and without

154   WORM-EATING WARBLER.

bars or markings. The lower parts are olive buff, lightest on the throat and belly, becoming grayish olive on the sides and flanks. The top of the

Worm-eating Warbler.

Helmitherus vermivorus (Gmel.).

head and sides of the face are brighter buff, and this color is intersected by four broad lines of clear black, two on top of the head from the base of the bill back, and one, passing from the bill back and through each eye. The birds are five inches and a half long. The sexes are alike.

The nest is built on the ground of leaves and plant fibres, and three to six white eggs are laid. They are spotted, with varying shades of brown, and are a little less than seven tenths of an inch long, and more than half an inch in their smaller diameter. The Worm-eating Warblers range throughout the Eastern United States, breeding where they occur, as far north as Southern New York and Connecticut, Southern Illinois and Nebraska. They winter in the West Indies, Central America, and Northern South America.

This Warbler is found in open dry woods and wooded hillsides. It is seen generally on or near the ground, and is slow and dignified in its movements, reminding one of a thrush in its ways. Not a common bird, it is by no means rare, but often escapes observation from its quiet retiring habits.

This so called Creeper is of all the Warblers the most easily recognized. The picture of a male bird on the opposite page is better than any descrip-

Black and White Warbler.

Mniotilta varia (Linn.).

tion. The colors of the bird are black and white. The female is similar, but has less black striping below, and the sides are washed with brownish. Young birds resemble the female, and often have a brownish suffusion above as well as below.

The birds occur, during their migrations, throughout Eastern North America, ranging as far north as Fort Simpson, and breed from Virginia and Kansas northward. They winter from the Gulf States south through the West Indies and Central America.

The nest is built on the ground, generally at the base of some tree or log, and is made of bark and plant fibres, lined with fine roots and hairs. Four or five eggs are laid. They are white, spotted with varying shades of brown, mainly about the larger end, and are nearly seven tenths of an inch long, and over half an inch in their smaller diameter.

140   BLACK AND WHITE WARBLER. ADULT MALE.

During the spring and fall migration the Solitary Vireo comes early and remains late, in the middle districts till almost the first of November. He is one of the few birds whose clear, musical song is heard at **Blue-headed Vireo.** both seasons.
Vireo solitarius (Wils.).
He is slaty blue on the top and sides of the head, shading into olive green on the back; the rump is brighter. There is a white *ring around the eye, and a white space between the eye and the bill.* The two white wing bars, and the lower parts are white, becoming yellowish on the sides and flanks. The sexes are alike, and the birds in the autumn are rather duller in general color, the blue of the head and the wing bars having olive and yellowish suffusions.

The nest is a hanging one, generally about ten feet from the ground, placed in the fork of a branch to which it is attached, except where its edge fills the space at the wider part of the angle of the fork. It is built of plant fibres and pine needles. The eggs, three or four in number, are white, sparsely spotted with dots of dark brown and black, mainly at the larger end. They are four fifths of an inch long, and nearly three fifths of an inch in the smaller diameter.

The birds range throughout Eastern North America, north to Hudson's Bay, during the migrations. They breed from Connecticut and the northern part of the Lake States northward. They winter from Florida southward to Central America.

The Mountain Solitary Vireo is a geographical race occupying the higher portions of the Alleghanies in Southwestern Virginia, and North and South Carolina during the breeding season. In the fall **Mountain Solitary** and winter it extends its migrations southward to Florida, **Vireo.** where it is of frequent occurrence during the colder parts
Vireo solitarius alticola
Brewst. of the year.

It is similar in plumage to the Solitary Vireo proper, but the upper parts are much darker and of a dusky, plumbeous tone on the back as well as on the head. The shading of greenish on the back is much fainter. The birds are larger, with comparatively much larger bills.

The nesting habits are essentially the same as those of the Solitary Vireo.

I found this bird a not uncommon summer resident, and breeding, during the summer of 1889, in the vicinity of Mountain Lake, Giles County, Virginia, at an altitude of from three to four thousand feet.

142   SOLITARY VIREO.

This is a western ally of the Solitary Vireo, and has been recorded once as an accidental visitor at Peterboro, New York. It is larger and is more

Plumbeous Vireo.
Vireo solitarius plumbeus
(Coues).

leaden gray above, including the rump; the sides and flanks are also more leaden, with little if any suffusion of yellowish or olive.

The Yellow-throated Vireo is the most brightly colored of the Vireos of Eastern North America. It is a robustly built, rather heavy bird, about five

Yellow-throated Vireo.
Vireo flavifrons Vieill.

inches and a half long. Its bright yellow throat and breast and a distinct yellow ring about the eye serve to identify it. The upper parts are clear olive green shading into grayish on the rump. · There are two white wing bars.

It builds a hanging nest of plant fibres, more or less covered on the outside with bits of lichen. This structure is suspended from a forked branch, generally more than twenty feet from the ground. The eggs are similar in color and markings to those of the Solitary Vireo. and vary from three to four in number. They are about four fifths of an inch long, and nearly three fifths of an inch wide.

The birds are found throughout Eastern North America, north to Ontario and Manitoba, and breed rather locally from Florida north through this territory. They winter in Central America and the Tropics.

This Vireo is a fine song bird and one of distinguished appearance. It frequents the higher branches of trees, in cultivated grounds about houses, as well as in the deeper woods.

This small Vireo, four inches and three quarters in length, is grayish olive green above, more markedly gray on the top of the head. It has a

Philadelphia Vireo.
Vireo philadelphicus (Cass.).

whitish stripe above the eye and the *entire lower parts are pale sulphur yellow*, most intense on the breast. The sexes are alike, and the yellow of the lower parts is more marked in individuals observed in the autumn. The nest is much like that of the Red-eyed Vireo and the eggs similar in color and markings.

The birds are found, during their migrations, throughout Eastern North America as far north as Hudson's Bay. They breed from Northern New England northward. They winter in Central America. Though common where they breed, they appear to be rare during the migrations, and seldom

YELLOW-THROATED VIREO.

come under the eye of the observer. They have not been recorded from the West Indies, and their route of travel is not well determined.

The song is said to be almost identical with that of the Red-eyed Vireo, and its general habits to be similar.

The Summer Tanager is about seven inches and a half long, and the male birds are light vermilion red throughout, brighter on the lower, duller on the upper parts. The wings and tail are dusky, edged **Summer Tanager.** with red. The female is plain olive, green above, and *Piranga rubra (Linn.).* saffron or yellowish buff beneath. Immature birds are like the females in general appearance.

The nest is made of rather coarse dry grass and plant stalks, is shallow and loosely put together, and is located near the end of the limb, usually about twenty feet from the ground. The eggs are greenish blue, profusely spotted all over, with brown markings of varying shades. They are a little over nine tenths of an inch long, and nearly seven tenths of an inch in width.

The birds are found in Eastern North America, north to Southern New Jersey and Illinois, and accidental occurrences have been recorded at more northern points. They breed throughout their regular range, and winter in Central America and Northern South America.

Tanagers, especially male birds, exhibit a marked amount of individuality in the shade and intensity of their gay colors ; and it is not uncommon to find individuals of a color and sometimes pattern far from that of typical birds. Thus male Summer Tanagers are, during the breeding season, occasionally bright yellow or pale rose color, and I have secured specimens that show the two colors mixed, when the birds seemed to be in full nuptial dress and not moulting.

This bird is rather smaller than the Summer Tanager, about seven inches long. The brilliant male in his intense scarlet coat, relieved in fine contrast by his lustrous black wings and tail, is familiar **Scarlet Tanager.** to all. The female is olive green above, with rather *Piranga erythromelas Vieill.* dusky greenish wings and tail, and the under parts are greenish yellow. Young males are like the female, but have blacker wings and tail. The adult male moults his scarlet coat in late August, and in the fall and winter is much like the female, save that the back is richer olive green, the lower parts are clearer yellow, and the wings and tail are clear black. The change from this phase of plumage to the bright breeding dress does not seem to have been observed, but is doubtless reached by a partial moult in the winter months.

SCARLET TANAGER. ADULT FEMALE.

The nest is much like that of the Summer Tanager, and a fairly typical structure is reproduced, from a photograph, on the succeeding page. The eggs are not quite as large as those of the Summer Tanager, but in color and markings are similar.

The birds are found as far north as Southern Ontario and Manitoba and breed from Virginia and Southern Illinois north. They winter in the same region as the Summer Tanager.

The males of the Scarlet Tanager vary much in the shade and intensity of both the red of the body and black of the wings and tail. They also present curious examples of color variation, similar to that referred to in the case of the Summer Tanager, viz., variety of color and pattern. One of the most frequent of these divergences is in the direction of one or two more or less clearly defined scarlet or bright yellow wing bars. These occur most often in very intensely colored birds.

147   NEST AND EGGS OF SCARLET TANAGER.

The male of this Tanager is a brilliantly colored yellow and black bird, with the head and neck rose red, most brilliant on the crown and back. The wings and tail are black. There are two conspicuous bright yellow wing bars. The females are very like those of the Scarlet Tanager, save that the wing bands are present, and are obscure yellow in color. It is a western bird, common in the Rocky Mountain region, as far north as British Columbia. It has been recorded from Massachusetts, Connecticut, and New York.

Louisiana Tanager.
Piranga ludoviciana (Wils.).

The Rose-breasted Grosbeak is known to most of us. The beautiful male bird with hood and back of pure black, his clear white under parts, with a striking patch of bright rose color on the breast. When flying close by he exhibits the same rose beneath his wings, and shows the clear white markings of wings, rump, and tail in strong contrast to the black. The female is like a large, robust sparrow, has a white line over the eye, and a buffy

Rose-breasted Grosbeak.
Zamelodia ludoviciana (Linn.).

ROSE-BREASTED GROSBEAK.
YOUNG MALE IN AUTUMN. LOWER PARTS.

ROSE-BREASTED GROSBEAK'S NEST.

median line on the head. Her general color is buffy brown throughout, and the wings have an area of light orange beneath, replacing the rose color of the male. This characteristic, with the size, head markings, and robust bill serve to make recognition an easy matter. Immature birds are like the female in general color, but the young males have the area under the wing and sometimes on the breast *rosy red*. The birds average rather more than eight inches long.

ROSE-BREASTED GROSBEAK.
MALE OF FIRST YEAR IN AUTUMN. UPPER PARTS

These are wood birds of the more open, well watered forests. In such woodlands where there is dense undergrowth, along some brook they find congenial localities, where they build their nests and rear their young.

The nest is a shallow, thin saucer of rootlets, twigs, and plant fibres, and is placed in dense bushy undergrowth or on the lower limbs of trees from five to twenty feet from the ground. A typical example is given in the accompanying illustration. The eggs, generally four in number, are rather light blue marked in a pattern, shown in the picture, with varying shades of brown. They are more than nine tenths of an inch long and a little less than seven tenths width.

The birds are found throughout Eastern North America as far north as

ROSE-BREASTED GROSBEAK. MALE IN SPRING.

Maine, Southern Canada, and Manitoba. They breed from Kansas and the higher parts of North Carolina and Virginia north. They winter in Central and South America. To conclude, they are noted songsters, are of great use in destroying harmful insects, and while more abundant generally in such localities as I have indicated, are frequently met with in cultivated grounds, and of late years are found commonly in some of our thickly built suburban

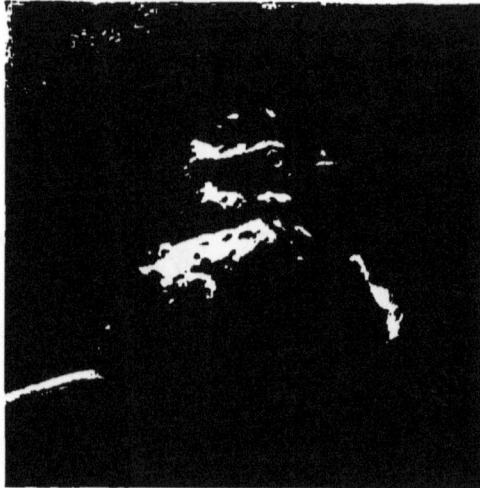

YOUNG FEMALE ROSE-BREASTED GROSBEAK.
PHOTOGRAPHED JUST AFTER LEAVING THE NEST. SOUTH ORANGE, NEW JERSEY, 12TH JUNE, 1897.

towns, where, undisturbed by the vicinity of man, they seem as much at home as in the wilder woodlands. Such conditions I have observed in the town of Cambridge, Massachusetts, where this is an almost abundant garden bird, and in South Orange, New Jersey, and vicinity much the same is the case. I have been able to obtain a series of photographs of two young female birds from the time of leaving the nest, and they are reproduced with notes on this and the following pages.

In this connection it will be of interest to both students and general readers to know that these two birds are at this time, January 1, 1898, still in the author's possession, both alive and strong healthy birds. They have

THE TWO BIRDS PHOTOGRAPHED TOGETHER.

IN THIS PICTURE THEY ARE TWO DAYS OLDER.

'54   ONE OF THE SAME BIRDS PHOTOGRAPHED FIVE DAYS LATER, ON JUNE 19TH.

passed through one partial and one complete moult and are in full plumage. They are extremely tame, and while not manifesting the inquisitive and investigating characteristics of Blue Jays and Baltimore Orioles, they betray certain kinds of attachment to different persons that is not so apparent in the two other kinds of birds. They are not caged, but are at liberty in a large room where they associate in the most friendly way with Thrushes, Robins, Cardinals, Chats, Song Sparrows, and Orioles.

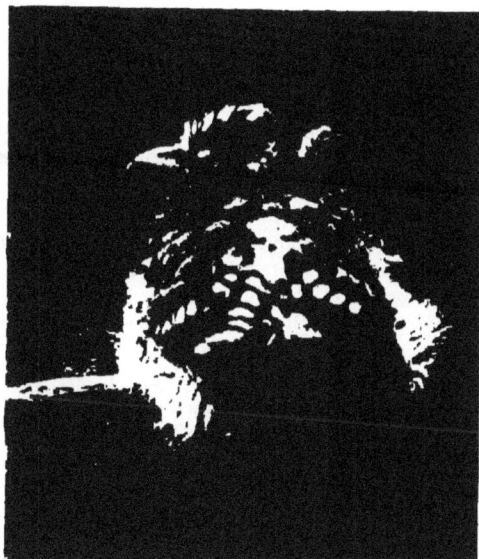

ANOTHER BIRD OF THE SAME BROOD, SHOWING THE BACK.
PHOTOGRAPHED ON JUNE 12, 1897.

The Black-headed Grosbeak is a bird, in form much like our Rose-breasted Grosbeak. It has a black head. The throat and a stripe behind the eye and on the crown are light buffy brown in color.

Black-headed Grosbeak.
Zamelodia melanocephala (Swains.).

The wings and tail are similar to those of the Rose-breasted Grosbeak. The color below the wings, and concealed by them when closed, is gamboge yellow or lemon.

This is a bird found from Middle Kansas to the Pacific Coast. They range north to British Columbia and Montana, and south into Mexico. There is the record of the capture of one in Michigan, where it must be regarded as accidental.

There is a large thrush-like looking bird that visits us in the early spring and late fall. You will notice it in thickets along the roadside, in brush heaps and tangles in the woods. Often only a pair,

Fox Sparrow.
Passerella iliaca (Merr.).

but frequently half a dozen or more of these birds associate together, and the company is often swelled to large proportions by Snowbirds and White-throated Sparrows.

The birds are about seven inches long, and of a general reddish brown tone. On the upper parts the underlying color is bluish slate, which shows plainly in the region back of the eye, more obscurely on the top of the head and back, and becomes entirely concealed on the rump, which is bright reddish brown. The tail is similar to the rump in shade, and the wings are dusky with reddish brown markings. The lower parts are white, heavily marked and spotted on the throat, breast, and chest, and streaked on the sides and flanks with brown. The belly, and the feathers under the tail, are white, the latter having in some individuals brownish streaking. The lower part of the bill is yellowish, and the eyes brown.

The Fox Sparrow breeds on the ground and in low bushes, building a nest of coarse plant fibres and grass, lined with finer material, hair and feathers. Three to five light blue eggs are laid. They are evenly marked, sometimes blotched, with varying shades of brown, and about nine tenths of an inch long, and a little over three fifths of an inch in width.

The birds are found throughout Eastern North America from the Arctic regions to the Gulf States. They breed from the Magdalen Islands and Manitoba northward, and winter chiefly south of the Potomac and Ohio Rivers.

The Piney-woods Sparrow is found in the grassy pine woods of Southern Georgia and Florida, where it breeds, and it moves but little to the south in

Pine-woods Sparrow.
Peucæa æstivalis (Licht.).

winter, returning early in the spring to its breeding grounds.

It is a bird in general build not unlike a Song Sparrow, and is about five inches and three quarters long. The feathers of the

FOX SPARROW.

upper parts are chestnut brown, with black streaks and gray edgings. There is a gray stripe over the eye and the bend of the wing shows yellow. The tail is long in proportion to the entire bird ; its feathers are narrow and the outer ones are much the shorter. The lower parts are dull whitish, the breast grayish brown, the sides and flanks darker ashy brown, and the belly lighter or white. The breast is sometimes marked with a few brown spots. The sexes are alike.

The nest is built on the ground, sometimes in a thick bunch of grasses, but more frequently under the low saw palmetto. The eggs, three to five in number, are white unmarked, a little over seven tenths of an inch long and three fifths of an inch in width.

This bird is a close ally of the last, as indicated by its name. It is a bird however of much wider distribution, being found in the Carolinas, Northern

Bachman's Sparrow.

Peucæa æstivalis bachmanii (Aud.),

Georgia, and Alabama, and west to Texas. In the warmer portions of the year, it occurs in the Mississippi Valley as far north as Southern Illinois and Indiana. It winters in the South Atlantic and Gulf States to Southern Florida.

Very similar to the Piney-woods Sparrow, this bird is much more tawny above, often without black markings, and brighter colored throughout. The stripe over the eye is buff in color, the breast and sides are distinctly shaded with the same color. The nesting and eggs are very similar in the two birds, though the Bachman's Finch is said to build a covered nest.

The White-throated Sparrow is one of our commonest finches during the migrations, and though introduced as a wood bird it is equally at home

White-throated Sparrow.

Zonotrichia albicollis (Gmel.),

in hedgerows, along roads, and in the thickets and shrubberies about the house.

The birds are about six inches and three quarters long, and are readily distinguished by their size, a yellow mark in front of the eye, the yellow on the bend of the wing, and the markings on the head and throat. In adult birds of both sexes there is a narrow white stripe in the centre of the crown, between two wider black stripes which join the yellow region above and in front of the eye. A white stripe proceeds from the eye backward. The back is brown, each feather streaked with black and finely bordered with buffy or grayish white. The rump is gray-

ish and generally unmarked.   The tail and wings are dusky, the latter with
the feathers edged with buffy brown and whitish.   The belly and feathers beneath the tail are white, the flanks, sides, and
breast slate gray.   There are usually traces
of fine *wavy, transverse, darker markings on
the breast*, the *throat patch* is of *pure white,
clearly defined* against *the color of the breast
and sides of the neck.*

The birds vary much with age and season, the white and black markings of the head
are often obscure, but traces of the yellow in
front of the eye and on the bend of the wing,
in connection with the more or less defined
white throat patch, will render recognition
an easy matter.

HEAD OF WHITE-THROATED SPAR-
ROW. ADULT.

They nest on the ground, or near it,
building a structure of coarse grasses and plant fibre lined with finer material.

WHITE-THROATED SPARROW. IMMATURE. AUTUMN PLUMAGE.

From three to five bluish white eggs are laid. These are evenly and often heavily marked with spots of varying shades of brown. They are about four fifths of an inch long and three fifths of an inch in width.

The birds are found throughout Eastern North America, as far north as Labrador and the Fur Countries. They breed from Northern New England, New York, and Michigan northward, and winter from Southern New England to Florida.

As songsters they have become celebrated for the sweet plaintive character of their notes. Though rarely heard during their migrations, their song is a feature of the regions where they breed. Here this sparrow is known and praised as the " Peabody-bird."

White-crowned Sparrow.

Zonotrichia leucophrys (Forst.).

This is a much rarer Sparrow in the east than its near relative, the White-throat. It comes much later in the spring, and remains a shorter time. It is a bird of about the same size and proportions, and of very similar habits, frequenting like localities. The adult birds are conspicuous among sparrows by their head markings. These are in brief a broad crown patch of white, with an equally broad band of black on each side of it, and a narrow white stripe from just over the eye, back along the lower edge of the broader black one. The back of the neck, the throat, and breast are light slate gray. This color shades into whiter on the belly, and buffy or brownish on the feathers beneath the tail, and on the flanks and sides. The back is brown, with a decidedly grayish tinge and slaty edgings to the feathers. The rump is plain ashy brown, the tail and wings dusky, the feathers of the latter edged with gray and whitish gray.

The sexes are alike, and the pattern is carried out in immature birds, the white areas of the crown becoming buffy, the black ones brown, and general brownish suffusion throughout, taking the place of the clear ashy gray of the adult.

These birds nest in a manner very similar to a White-throated Sparrow, and their eggs differ in being more greenish blue in color, spotted with brighter shades of brown, and are a little larger.

The White-crowned Sparrow is found in North America at large, but is more common west of the Alleghanies. It breeds in Eastern North America, north of the United States to Labrador and the Fur Countries. It winters from Mexico southward.

The Pine Finch is a small bird about five inches long. Its ground color is gray or grayish brown on the upper and whitish gray on the lower parts.

Pine Siskin.
Spinus pinus (Wils.).
Both regions are streaked throughout with dusky brown or blackish markings. The wing feathers are *yellow at their bases*, and are edged with yellowish olive. The tail is dusky and all but the middle feathers are *yellow at their bases*. The bill is conical and acutely pointed. The sexes are alike, the young are similar, and in all plumages the yellow markings obtain.

The birds nest in evergreen trees, building a structure of fine roots and plant fibres, lined with plant down, hair, and fibre. The eggs resemble those of a Goldfinch in color, but are sparingly spotted with varying shades of brown. They are a little more than three fifths of an inch long, and half an inch in width.

The birds are found in North America at large, they breed chiefly north of the United States, though there are records from New York State. They winter south to the South Atlantic and Gulf States.

These little Finches are gregarious, at least during their migrations, and are in habit similar to the Goldfinch, but feed more in pines and kindred trees.

There are two kinds of Crossbills to be found in Eastern North America. They are both readily recognized by their curved bills, crossing at the tips.

American Crossbill.
Loxia curvirostra minor
(Brehm).
The Red or American Crossbill is a little the larger, being six inches and one fifth in length, whereas the White-winged Crossbill is about six inches long. Adult males of both species are reddish, the Red Crossbill being deep brick red, and its ally rosy red in tint. The females of both are olive green with gray-

White-winged Crossbill.
Loxia leucoptera Gmel.
ish suffusions, and the young of both are gray streaked birds in general appearance. But in all plumages the wings of the Red Crossbill *are unmarked*, while the wings of the White-winged Crossbill always *show white markings*.

Their breeding habits are similar, both constructing their nests of plant fibre and moss, in pine and other evergreen trees, about twenty feet from the ground. The eggs of both are pale bluish green, spotted with varying shades of brown. They are from three to four in number, about three quarters of an inch long, and rather more than half an inch in width.

The Red Crossbill is irregularly resident in the Alleghanies, in localities

as far south as North Carolina, and regularly from the Northern States northward. It is an erratic migraht during the colder parts of the year, appearing
at times in the extreme Southern States.

The White-winged Crossbill is a more northern bird and breeds, so far
as known, in the northern parts of North America, migrating south into the
United States, in winter as far south as Illinois and Virginia.

AMERICAN CROSSBILL.

Both kinds of Crossbills are gregarious, appearing in flocks of varying
size. They resemble the parrots in their habit of using their curious bill in
climbing about in search of the coniferous and other seeds that form their
food supply.

The Red Crossbill is a close ally of the European bird of like name, and
is recognized by naturalists as a sub-species or geographical race of that bird.

The Pine Grosbeak is found throughout the boreal regions of the
Northern Hemisphere and breeds in high latitudes. To us it is known as a
winter visitor in the more northern of the United States,
Pine Grosbeak. rarely being recorded south of New Jersey and Pennsyl-
Pinicola enucleator (Linn.). vania.

It is a large robust bird, about nine inches long, a typical finch in build
and contour. The adult male is in general color rosy red over an under
coloring of slaty gray, the red prevailing on the crown, rump, and breast, and

the gray showing on the shoulders, sides, flanks, belly, and beneath the tail. The wings are dusky, with whitish edgings becoming pinkish in some individuals, and the tail is dusky. Adult females have the wings and tail similar to the adult male, but are otherwise grayish slate color, with a washing of yellowish olive, where the rosy red prevails in the male. Immature birds are similar in general appearance to the female.

But little is known of the life history and habits of these Grosbeaks at their breeding grounds, but the nest is built of rather coarse plant fibres and twigs, and lined with finer material, in pine and kindred trees, not far from the ground. The eggs are a little over an inch long, and nearly three quarters of an inch in their other diameter. They are greenish blue in color, spotted with varying shades of brown.

Another bird of boreal distribution, occupying the interior portions of Northern North America from Manitoba northward, and migrating regularly **Evening Grosbeak.** into the upper Mississippi Valley, and erratically to the **Coccothraustes vespertinus** North Atlantic States, is the Evening Grosbeak. **(Coop.).** A robustly built finch, some eight inches in length, with a particularly large and heavy bill, it is among the most striking of the family, both on account of its fine appearance, the texture of its plumage, and its beautiful coloring.

The male has a yellow forehead and black crown. The belly and tail are black. The sides of the head, the neck, and back are brownish olive, changing to yellow on the shoulders and rump. The wings are in the main black, but there is a considerable area of white below the yellow on the shoulders and some of the longer feathers are tipped with white. The region under the wings is yellow.

The females and immature birds are grayish brown, with yellowish suffusion especially on the back of the head and neck, lighter on the breast and belly. The tail is black, tipped with white on the inner webs of the feathers, and there is a whitish patch at the base of some of the longer wing feathers.

But little is known of the summer habits of this bird, and its very appearances where it occurs more commonly in winter are erratic and nomadic. At points in Minnesota, Wisconsin, and in Northern Illinois and Iowa, it seems to be of more regular occurrence than elsewhere. The records of its comings and goings embrace all months from September to late in May. The birds at such times are singularly tame and frequent the vicinity of houses

and trees in the streets of towns, as well as the woodland, feeding on berries
of various kinds.

The Florida Crow is closely allied to the Common Crow of the more
northern States. It is almost identical in general appearance, but is rather
Florida Crow.   longer, with tail and wings proportionately smaller.
Corvus americanus floridanus   The nesting habits and eggs are practically the same
Baird.   as those of the more widely distributed bird. This bird
is found in the pine woods of Florida, and does not like its ally affect open
fields and cultivated grounds.

The Raven is represented in the bird fauna of Eastern North America by
a geographical race known as the Northern Raven. It is an ally of the Old
Northern Raven. World bird as indicated by its scientific name. In general
Corvus corax principalis color it is like the Common Crow, black with a deep steel
Ridgw.   blue sheen to the plumage. The feathers on the throat
and sides of the neck are what would be called "hackles" in the barnyard
fowl. That is they are long and narrow and pointed at the end. Further
they have the appearance of being separate from one another. These characteristic feathers and the size of the bird, rather more than twenty-two
inches long, will serve to identify it.

The Raven nests in tall trees, or on some shelf on the face of a cliff,
building much after the manner of a crow, and repairing the same nest year
after year, if undisturbed. The eggs vary in number, from two to seven.
They also vary in ground color, from bluish green to light olive. They are
marked much as are the Crow's eggs with spots and splashes of olive and
darker brown. They are nearly two inches long and about an inch and a
third broad.

In Eastern North America the Raven is found more commonly north of
the United States. In the Eastern United States its distribution is local
from Northern Michigan and Maine, southward to Virginia and North Carolina, in the mountains.

The birds also occur in New Jersey, where the vast cedar swamps near
the coast are still frequented by a limited number. I have seen and obtained
specimens in this region, as well as in Asheville, North Carolina, where they
resorted to the vicinity of slaughter houses, outside of the town. Here I

secured five of these birds in the fall of 1889, and they were rather common. During the summer of the same year, spent at Mountain Lake, Giles County, Virginia, I frequently saw Ravens.

In the great unbroken forests of Northern New York and Maine, and northward from these points and from Northern Minnesota to the Fur Countries, there is a large and conspicuous bird, known as the **Canada Jay.** Canada Jay, or Whiskey Jack. In form it is much like Perisoreus canadensis (Linn.). an exaggerated Titmouse, and has the same hair-like, downy plumage, so characteristic of the Chickadee.

The bird is about a foot long, and the general color is gray, relieved by white on the forward parts of the head, the throat, and sides of the neck, in marked contrast to the back of the head and neck, which are black. The back, wings, and tail are gray with whitish suffusion, formed by the white tips of each feather. The chest, belly, and sides are slaty gray.

The sexes are alike, and the immature are similar to the adults, but the gray is more obscure, and the markings of the head are not clearly defined.

The birds breed generally in pine trees, building a bulky nest of coarse roots and plant fibres much like that of the Blue Jay. The eggs, usually four in number, are whitish thickly speckled with gray browns of several shades. They are about one inch and an eighth long, and four fifths of an inch in width.

One of the characteristic and abundant birds of the regions it occupies, this bird has become familiar to hunters and fishermen, and to the lumbermen of the northern pine woods, by its sociable habits and its propensity for pilfering about the camp.

The sea-coast of Labrador is the region where this ally of the Canada Jay is found. The Labrador Jay is like the Canada in general appearance, **Labrador Jay.** but the black coloring of the head and back of the neck Perisoreus canadensis nigri- is more extensive, reaching to and surrounding the recapillus Ridgw. gion about the eye, and the entire bird is darker in color.

The Scrub or Florida Jay is rather a long, slim bird of a wren-like character, living in thick, dense tangles of low growth known locally as "scrub."

The bird is about eleven inches long. Its general colors are blue, gray, and white. The blue is found on the top and sides of the head and neck, and on the wings and tail. The under parts are white

Florida Jay.
Aphelocoma floridana (Bartr.).

or grayish white. There are faint streakings of gray blue on the throat and breast, in the latter region forming a more or less distinct band. The back is grayish brown, clearly contrasted and defined by the surrounding colors.

These Jays are locally distributed, and are practically limited to the so-called "scrubs," where they breed and raise their young.

The nest is rather loosely and carelessly put together, is composed of sticks and coarse weed stalks, and lined with finer material of like nature. The eggs are generally four in number and are yellowish green in color, much spotted with varying shades of brown from olive to umber. They are about one inch and one tenth long, and four fifths of an inch wide.

This bird is found in the Eastern United States as far north as Southern New York, Connecticut, and Southern Michigan. It breeds from

Green-crested Flycatcher.
Empidonax virescens (Vieill.).

Florida northward, and finds a winter home in Central America. It is the next to the largest of this group of Flycatchers, the Alder Flycatcher being the largest. It is brighter olive green above than the Yellow-bellied Flycatcher. The wings and tail are dusky, with edgings like the back. There are two whitish wing bars, having a yellow tinge. The under parts are white, generally with a faint yellowish tone suffused with green. *The throat is always white.*.

The birds are about five inches and three quarters long. The nest is usually built in a fork of a branch, about ten feet from the ground. It is composed of plant fibres compactly built together, and is, in general character, a shallow saucer. The eggs vary from two to four in number, are white in color, with brown spots about the larger end. They are a little larger than those of the Yellow-bellied Flycatcher.

These Flycatchers frequent open woods where there are small brooks, and are generally to be found perched in the lower branches of the trees.

This Flycatcher is a little larger than the Least Flycatcher and smaller than the Green-crested Flycatcher. It is deep olive green above, with tail and wings dusky. The tail is edged with olive green. There are two yellow bars on each wing. The under parts are light greenish yellow, brightest on the *belly*, and obscured by olive green on the *breast* and *upper throat. The throat is always yellow.* The bird is about five inches and three fifths in length.

Yellow-bellied Flycatcher.

Empidonax flaviventris
Baird.

The nest is placed on the ground, sometimes under a tree root, and again sunken in moss. It is made outwardly of moss and lined with fine grasses and plant fibres. The four eggs are white, or pale buff, with many pale reddish brown markings, more numerous at the larger end. They are two thirds of an inch long, and a little over half an inch wide.

YELLOW-BELLIED FLYCATCHER.

The Yellow-bellied Flycatcher is found in Eastern North America, as far north as Southern Labrador. It breeds from the Northern United States northward through its range, and winters in Eastern Mexico and Central America.

This is one of the earliest migrant Flycatchers to appear in the autumn, as it comes back from its breeding grounds, south to New York and New Jersey, in August. It seeks at all times dark and shady places in the woodlands.

The Wood Pewee, while essentially a wood bird, frequents cultivated grounds, especially lawns well shaded by large trees. Old apple orchards are also favorite hunting and breeding grounds for these birds.

Bird Studies.

They are about six inches and a half long. The upper parts are decided olive brown, shaded sometimes with grayish olive green. The **Wood Pewee.** wings and tail are darker, and there are two whitish Contopus virens (Linn.). wing bars. The lower parts are grayish white, sometimes faintly yellow, becoming olive gray on the breast, sides, and flanks. The sexes are alike, and immature birds are much the same in appearance, except that the wing bars are brownish buff, and there are suggestions of yellowish on the lower parts.

WOOD PEWEE.

The nest is placed on a horizontal limb, from fifteen to fifty feet from the ground. It is a well built structure of fine grasses, plant fibres, and moss, covered with lichens on the outside. The three or four white eggs are marked with a circle of brown spots of varying shades at the larger end. They are about seven tenths of an inch long, and more than half an inch in their other diameter.

The birds are found in Eastern North America, during the warmer portions of the year ranging as far north as Newfoundland, and breeding from Florida northward. They winter in Central America.

In the Woods.

This is a robust, strongly built bird, about seven inches and a half long. It is conspicuous from its size and its habit of perching on some high dead tree, or limb projecting from the live branches below. In such places, a miniature hawk in pose and appearance, he watches for his passing insect prey, which he seizes on the wing, returning to the same post to resume his guard and consume his captive.

Olive-sided Flycatcher.
Contopus borealis (Swains.).

The upper parts are dusky olive. The throat and belly are white or yellowish white, connected generally by a narrow stripe of the same color down the centre of the breast and chest. From these regions the remainder of the lower parts is about the same color as the back, except for a tuft of white feathers on either side of and below the rump.

The sexes are alike, and immature birds are similar to the adults, except that they are more olive above, yellower beneath, and the feathers of the shoulders are edged with buff.

The birds nest usually in pine or kindred trees, building a structure of moss, plant fibres, and twigs, near the end of a limb. Three to five eggs are laid, creamy buff in color, spotted about the larger end with varying shades of brown. They are rather more than four fifths of an inch long, and three fifths of an inch in their smaller diameter.

The birds are found throughout North America, during the warmer parts of the year, as far north at least as British Columbia. They breed from Massachusetts and Minnesota northward, and on the higher mountain ranges south to North Carolina. They winter in Central and South America.

A Flycatcher that breeds in holes in trees, that is alike conspicuous for his size, fine colors, and efforts at vocalism, is surely a bird to attract attention. Such a one is the Great-crested Flycatcher. He is nine inches long. The top of the head, back of the neck, and the back are greenish olive brown. The wings and tail are darker. Some of the larger feathers of the wing are edged with cinnamon and the inner webs of all but the two middle tail feathers are the same color, but rather lighter. The throat, region in front of the eye, breast, and sides of the neck are bright slate gray. The rest of the lower parts are clear sulphur yellow.

Crested Flycatcher.
Myiarchus crinitus (Linn.).

The sexes are alike and immature birds are similar, but generally more olive.

170   GREAT-CRESTED FLYCATCHER.

The nest is placed in some deserted woodpecker's hole or natural cavity in a dead limb. Here some twigs, grasses, and often pieces of cast off snake skin and a few feathers form a bed for the eggs or young. From three to six eggs are laid. They are buffy white in color, heavily lined and penciled all over with dark shades of brown. They are nine tenths of an inch long, and nearly seven tenths of an inch in width. These birds are found in the Eastern United States as far north as Southern Canada and New Brunswick. They breed from Florida northward throughout their range, and winter from South Florida southward to Central America.

The Whip-poor-will seems like a moth among birds. The quality of the feathers, their softness in color and markings, the very uncertainty of the flight when one is started from its bed of dry leaves **Whip-poor-will.** in some dusky part of a deep wood, all join to heighten Antrostomus vociferus (Wils.). the impression. The crepuscular habits of the bird form another link that adds emphasis to this fancy. After the sun is down and the dusk steals over the landscape, and again just before sunrise, and in the uncertain moonlight the Whip-poor-will leaving his day resorts seeks his food on the wing, hovering low over the edges of fields and even along highways. His excursions are not long, and between them he perches on some bare stone, or log, even on some fence post or rail, or perhaps on a stone wall, and with a single preliminary sudden "chuck" as an overture, begins his concert of Whip-poor-will, which has given him his name.

WHIP-POOR-WILL'S FOOT. COMB ON LONGEST CLAW.

The bird is about ten inches long, and of a general underlying color of dark brown. On each side of the top of the head this color is mottled with fine black and white markings, the centre of the crown being darker. The general color of the back is similar, but is covered and almost hidden when the wings are closed. On the shoulders areas of light buff and dark brown mottling alternate. The wings are dark brown with reddish bars. The tail

WHIP-POOR-WILL.

is much like the back, the three outer feathers being conspicuously white on their terminal half. There is a white band dividing the throat and breast. The throat is dark brown with traces of lighter brown mottling. The rest of the lower parts are of a light yellowish brown obscurely barred with dark brown.

The mouth is very large, and along its upper edge, from below the eye to the nostrils, stretches a row of long stiff hair-like bristles. The middle toe has a conspicuous *comb* on its claw.

The female is similar to the male, but the white areas on the tail and across the lower throat are replaced by buffy markings not so extensive as are the white areas in the male. Two eggs are laid on the ground, generally where there are dry dead leaves. They are dull white with lilac suffusions and some distinct dots of varying shades of brown. They are about one inch and an eighth long and more than four fifths of an inch in their other diameter.

The birds are found in Eastern North America as far north as New Brunswick and Manitoba. They breed from the Carolinas northward and winter from Florida southward.

There is a prototype of our Whip-poor-will taking that bird's place during the warmer portions of the year in our more southern districts. It is

Chuck-will's-widow.
Antrostomus carolinensis (Gmel.).

a larger bird, fully twelve inches long, and *lighter* in general color. The same character of mottling in more tawny colors prevails, and the white areas in the male are replaced by buffy in the female, on tail and throat. The white on the tail extends to more feathers and the throat band is more broken. The same bill bristles are found, but differ from those of the Whip-poor-will in having branches on their basal halves. This is the Chuck-will's-widow, whose song is much the same in quality as that of the Whip-poor-will, but has an added syllable and is much more deliberately uttered.

The method of nesting is similar in the two birds, as are the eggs, though those of the Chuck-will's-widow are larger, being one inch and two-fifths in length and about an inch in width.

The birds are found in the South Atlantic and Gulf States, from North Carolina southward, going as far north in the interior as Southern Illinois and Kansas. They breed throughout their United States range and winter from our southern boundary southward.

FLICKER OR GOLDEN-WINGED WOODPECKER. ADULT MALE.

ENTRANCE TO NEST OF FLICKER.

The Flicker is a harlequin among birds, with black and red, white and gold, exhibited as he passes by. You will know the bird by the patch of white on

Flicker.
Colaptes auratus (Linn.).

the rump and the yellow showing beneath the wings and tail in his flight. A closer look will reveal a clear black band across the breast, and an equally well defined black patch starting at the angle of the jaw, and passing back on the face to the region below the ear. There is a scarlet band on the back of the head. The top of the head is ashy gray, as is the back of the neck. The back is grayish brown barred with dark brown or black. The wings and tail are black or dusky seen from above, with yellow shafts to the feathers. Seen from below the wings and basal two thirds of the tail feathers are bright yellow or golden. The tips of the tail feathers are black. The throat and the sides of the face are a warm light brown, almost pinkish ; this color prevails to the black throat band. Back of the throat band the lower parts are whitish with suggestions of the color of the throat, and thickly marked with round black spots. The female is like the male, but lacks the black markings on the sides of the face below the eye. The birds are about a foot in length.

Like all woodpeckers, the Flicker breeds in holes in trees, generally, but not always, of their own excavation. The eggs, four to nine or ten in number, are pure white, and are an inch and one tenth in length, and a little less than nine tenths of an inch in width.

The birds are found in Eastern North America north to Hudson's Bay. They breed throughout their range and winter south from Illinois and Massachusetts.

Though true woodpeckers, they are often seen on the ground, where worms and larvæ of insects tempt them, and the smaller wild fruits of the dogwood, gumberry, and cherry are eagerly sought after in their season.

The Red-bellied Woodpecker is about nine inches and a half long. It impresses one, in a way, as a small Flicker, having a certain general resem-

Red-bellied Woodpecker.
Melanerpes carolinus (Linn.).

blance to that bird. In the adult male the entire top of the head, extending well back on the neck, is clear shining scarlet. The back is barred black and white. The feathers above the base of the tail are white, marked with black streaks. The wings are barred black and white, except that the larger feathers are black at their ends. The tail is marked in the same way, except on the two middle feathers, which have less white than the others. The sides of

177

FLICKER NEST.
SECTION SHOWING INTERIOR WITH EGGS

the head and under parts throughout are dull whitish ash, with the middle of the belly and the region below the, bill and often the breast tinged or suffused with scarlet. The female is like the male except that the scarlet of the upper parts is confined to the back of the head and neck and region about the nostrils, the crown being ashy gray.

The nest is built in holes in trees excavated by the birds, where four to six white eggs are laid. These are about an inch long and a little less than three quarters of an inch in their other diameter.

The birds are found in Eastern North America north regularly to Maryland on the coast, and to Dakota and Ontario in the interior. They breed from Florida throughout their range, and winter from Virginia and Ohio south.

When great unbroken forests clothed what are now the farming and agricultural lands of the Eastern United States ; when Manhattan Island **Pileated Wood-** was wooded and the Indians traded their beaver skins **pecker.** captured in the streams of Westchester County with the Ceophlœus pileatus (Linn.). Dutchmen in the village of New Amsterdam ; when the buffalo ranged into Ohio and even Western New York, probably on upper Broadway the woods were inhabited by a very large black woodpecker, known as the Pileated Woodpecker or Logcock.

But the Indians and buffaloes and Dutchmen are almost traditions now, and the great Woodpecker of the primeval forest has disappeared with their felling and is now practically restricted to the more remote localities where such forests still obtain. There he still is a comparatively common bird, be it in Florida or Maine.

Nearly a foot and a half long, of a general dusky black color, the male has a bright scarlet crown, with the feathers at the back of the head elongated into a pointed crest. There is a white border line to this crest separating it from the dusky region back of the eye. Another line begins at the nostrils and extends well down on the sides of the neck. It is of a general yellowish tinge at its starting point, becoming clear white on the sides of the face and neck. The lower parts are dusky black, the feathers being more or less tipped with yellowish white. The throat is white or yellowish white bordered by a scarlet patch extending back from the lower part of the bill. The wing feathers are white at their bases.

The female is similar to the male. The red of the head however is limited to the crest and there is no red patch at the base of the bill.

The nesting is like that of other woodpeckers, a hole excavated by the birds in some tree. One that I took in Florida was in a palmetto forty feet from the ground; another in a dead pine scarcely fifteen feet up. These were both found early in April, one containing four and the other five fresh eggs. They were white and about an inch and three tenths long and nearly an inch in their smaller diameter.

The yellow-bellied Woodpecker is decidedly a woodland bird during the breeding season, but during its migrations, particularly in fall, many appear

Yellow-bellied Woodpecker.

Sphyrapicus varius (Linn.).

in the orchards and trees about country houses. The male is a distinguished bird, with the top of head and his throat deep scarlet. The back is barred with yellowish white on a black ground. There is a large white area on

YELLOW-BELLIED WOODPECKER.

each shoulder and the feathers of the wings are much marked with white. The general color of the tail is black, with white margins to the outer feathers. The breast is clear shiny black. Back of this the under parts are yellow, streaked with black and dusky on the sides. The adult female has a *white throat*, sometimes a scarlet crown, though this region is frequently clear black and again black with a mixture of scarlet feathers. Otherwise she is like the adult male. Immature birds resemble the female, but are duller colored generally and have brown or dusky brown on the top of the head, often with a mixture of dull red feathers.   Frequently there is an admixture of scarlet with the dull throat feathers.

The nesting is similar to that of most woodpeckers, a hole in some tree, excavated by the birds at varying heights from the ground.   Four to seven white eggs are laid which are nearly nine tenths of an inch long, and less than seven tenths of an inch in their other diameter.   The birds are migratory, and are found in Eastern North America as far north as Fort Simpson. They breed regularly from Northern New England north.   I found two pair breeding on the summits of the Alleghanies in Southwestern Virginia.   They winter from Virginia southward through the West Indies to Central America.

There are two kinds of Three-toed Woodpeckers in Eastern North America.   Their names indicate their characteristic feet.   Two toes are in front and one behind in both kinds.   They are both boreal birds, going as far north as the tree limit, and ranging south to the Northern United States.

The American Three-toed Woodpecker is limited in its distribution to the region east of the Rocky Mountains, while the Arctic Three-toed Wood-

American Three-toed Woodpecker.
Picoides americanus Brehm.
pecker occurs throughout Northern North America.   The smaller of the two is the American, which is about eight inches and three quarters long.   In the male the top of the head is black, spotted with white, and there is a yellow crown patch. The back is barred black and white, and the wings are black spotted with white.   The middle tail feathers are black, and the outer ones are black and white.   The under parts are white, the sides and flanks are barred with black. The female is similar, but lacks the yellow crown patch.

The Arctic Three-toed Woodpecker is larger, being nine and a half inches long.   The entire top of the head, except a yellow median crown

Arctic Three-toed Woodpecker.
Picoides arcticus (Swains.).
patch, and the back are a *clear lustrous black.*   There is a white line passing from the nostrils to below the eye. The other parts are colored much the same as in the smaller ally.   The female is like the male, but lacks the yellow crown patch. Both kinds nest near the ground in trees, in the characteristic manner of most woodpeckers.

The Red-cockaded Woodpecker, belying its scientific name, *borealis*, is a bird of Southern distribution, and is resident in the United States from

In the Woods.

North Carolina and Indian Territory south to Eastern Texas and through
the Gulf States. It is a bird of the pine forests, though found at times

Red-cockaded Woodpecker.
Dryobates borealis (Vieill.).

in other kinds of woods. One of the smaller woodpeckers,
it is about eight inches long. The male is clear black
on the top of the head, with a small and partly concealed
scarlet tuft of feathers on each side of the head above and back of the ear.
The region about the ear and on the sides of head is *white*, reminding one of
the similar markings of the Chickadee. This is separated from the throat by
a black line reaching from the bill to the shoulders. The back is barred with
black and white, and the wings are black spotted with white. The middle
tail feathers are black and the outer ones black and white. The under parts
are white streaked on the sides, flanks, and on the feathers below the tail
with dusky black. The female *lacks* the *scarlet tufts* on the head.

The hole for the nest is generally excavated high up in a pine in some
dead limb or branch.

The eggs are white, generally four in number, and about nine tenths
of an inch long, and nearly seven tenths of an inch in their smaller
diameter.

The Hairy is an exaggerated Downy Woodpecker, half as large again
as that bird, being about nine inches long. The colors and pattern are prac-

Hairy Woodpecker.
Dryobates villosus (Linn.).

tically the same, and sometimes a Downy Woodpecker
deceives one for a moment when his ruffled feathers add
to his size.

But the Hairy Woodpecker is eminently a woodland bird, and where
much timber has been cut off becomes noticeably rare, whereas his smaller
reflection seems really more at home in the trees about the houses than in
the forest.

The Hairy Woodpecker breeds generally in a dead tree, laying four or
five white eggs rather more than nine tenths of an inch long, and about seven
tenths of an inch in their other diameter. These birds are found in the
Eastern United States, from their northern border south to the Carolinas,
and breed throughout their range.

From the Carolinas south the Hairy Woodpecker is represented by a
closely allied geographical race. This Southern Hairy Woodpecker is

HAIRY WOODPECKER.

Southern Hairy Woodpecker.
Dryobates villosus audubonii (Swains.).

smaller, being but little over eight inches long, and is marked like its congener, except that there are fewer white spots on the wings and the white on the upper parts is more restricted.

Northern Hairy Woodpecker.
Dryobates villosus leucomelas (Bodd.).

Contrasted with this Southern type is the bird known as the Northern Hairy Woodpecker, which attains a maximum length of nearly eleven inches and is the representative geographical race north of the United States, sometimes crossing our northern borders in the colder portions of the year.

Yellow-billed Cuckoo.
Coccyzus americanus (Linn.).

Black-billed Cuckoo.
Coccyzus erythrophthalmus (Wils.).

There are two kinds of Cuckoos that are found in our woodlands, the Yellow-billed and Black-billed Cuckoos. They are both long slim birds of a general dove colored brown with slight greenish iridescence above, and grayish white below. The Black-bill is the smaller of the two, being about eleven inches and a half long, while the Yellow-billed bird is twelve inches or more in length. The Black-billed Cuckoo is distinguished by the color of its bill, indicated by its name. The other bird has the lower half of the bill yellow, except at the tip, and all but a few of the wing feathers are bright cinnamon brown on their inner webs, becoming darker brown at their tips. The two middle tail feathers are dove brown much darker than the back. The other tail feathers *are black*, the outer ones edged on the outer web and tipped with pure white, and all the others except the two middle ones tipped with white. The tail of the Black-billed Cuckoo is like the wings in color and the white tips and markings are not so extensive as in the Yellow-billed Cuckoo.

Both birds build nests, consisting of a loose platform of sticks and sometimes some grasses or other plant fibres. They are placed in the lower branches of the smaller trees or in bushes from five to ten feet from the ground. The eggs of the Yellow-billed Cuckoo vary from two to five in number, and are light greenish blue in color. They are about an inch and a fifth long and nine tenths of an inch in their smaller diameter. The Black-billed Cuckoo's eggs are rather darker in shade and a little smaller than those of its congener.

Both birds range throughout Eastern North America during the warmer

184   YELLOW-BILLED CUCKOO.

portions of the year. The Yellow-billed Cuckoo is found as far north as New Brunswick, Canada, and Minnesota and breeds throughout its North American range, wintering in Central America. The Black-billed Cuckoo breeds as far north as Labrador and Manitoba and winters in Central America and South America.

Both kinds of Cuckoos are great insect hunters, destroying multitudes of such caterpillars as infest our fruit and shade trees. While both are essentially woodland birds they also frequent the shade trees and orchards and are conspicuous when flying, and by their peculiar notes. But when in a tree few birds are better concealed by their color and quiet method of working.

The Mangrove Cuckoo is a bird of the West Indies that is of regular occurrence but not at all common on the Florida Keys and the southwest coast of that State. It is rather larger than the Yellow-billed Cuckoo and somewhat more heavily built than that bird.

Mangrove Cuckoo.
Coccyzus minor (Gmel.).

Of a similar color above, the feathers are more glossed. Below however these birds are deep buff, sometimes almost tawny, becoming lighter on the belly and on the feathers below the tail. The middle tail feathers are like the back, the rest being *black* with a broad area of *white* at their ends. The region about the ears is dusky or blackish. The bill is similarly colored to that of the Yellow-billed Cuckoo.

The building, breeding, and eggs are almost identical in the two birds.

Maynard's Cuckoo is a bird allied to the Mangrove Cuckoo, and is found, so far as known, in the Bahamas and on the Keys of the southern coast of Florida.

Maynard's Cuckoo.
Coccyzus minor maynardi (Ridgw.).

It is a bird about the same size as the Mangrove Cuckoo, but has the under parts pale gray or ashy white on the throat and breast, usually faintly washed with pale buff on the sides and belly. The bill is not so robust. These birds are of regular occurrence and much more common than the Mangrove Cuckoo in Florida.

This is a black cuckoo with a curiously *thin* bill, compressed laterally and therefore *high* from the bottom to the top. It is a common bird in the Bahamas and West Indies, where it is known as the "Tick Bird," and in places where cattle are grazed many of these birds follow after them or light on their backs in quest of parasites, which form their food. The birds are slim in build and have very long tails. The average length of the Ani is about fourteen inches.

Ani.
Crotophaga ani Linn.

When North America was settled by the Puritans in New England and the Cavaliers in Virginia, there were regularly in the latter State and as far north as Maryland, the Great Lakes, Iowa, and Nebraska, and sometimes in Pennsylvania and New York, what the early settlers and historians called *Parrots*.

Carolina Paroquet.
Conurus carolinensis (Linn.).

For formerly the Carolina Paroquet occupied all the region as far north as defined, and extended its range west to Colorado. So far as now known, the birds are found only in a few localities in the Indian Territory and in parts of Florida. Even the last twenty-five years has shown a marked decrease in the birds in Florida. During the winter of 1875 and 1876, which I spent in Sumpter County on Panasoffkee Lake, I saw large flocks of these birds daily, and also noticed many flocks the same year in passing up and down the Ocklawaha River. In the winter of 1879 and 1880 I saw only two flocks, and these were small. During the years from 1885 to 1892 I was in Florida for at least half of each year, and in 1886 the entire year, but I find only one record of Paroquets in these years, a small flock seen at a place called Linden. As my travels were taken to study the birds of the region, and as they extended to every point of the Gulf Coast from Cedar Keys to the Dry Tortugas, and also well into the interior, I conclude that but a small remnant of these birds exists, and while doubtless man is directly responsible for their actual destruction, yet I cannot but believe that the more subtle indirect influences growing out of the settlement of the country are far more responsible for the results than the actual slaughter, great as it has been, which has occurred.

The birds are about thirteen inches long. The adults have yellow heads, the color extending to the neck and throat, thus forming a hood of gold. This color is more intense, becoming orange or even reddish on the forehead and sides of the face. The bend of the wing and the feathers nearest to the feet are also yellow or orange, and the under surface of the tail is yellowish. The other portions of the body are bright grass green. Immature birds lack the yellow hood, the head being green, with more or less yellow or orange on the forehead and in front of the eyes.

It is rather noteworthy that we have no exact information as to the nesting habits of these birds; some accounts speak of them breeding in hollow trees, and others of their building nests on branches.

The eggs are white, and are one inch and a third long, by about an inch and a tenth in their other diameter.

The whole subject is well worthy of attention, so that we may have definite knowledge in regard to their nesting and breeding before the disappearance of the Paroquet, which now seems imminent, renders such investigation impossible.

Great Horned Owl.
Bubo virginianus (Gmel.).

This is the only large sized owl with pronounced ear tufts that we have in Eastern United States. The males are about twenty-two inches in length and the female, which is noticeably larger, often is over two feet long. The upper parts have a general brown tone produced by a fine mottling of various shades of ochre, brown, and black. The discs about the yellow eyes are similar in color but lighter in shade. There is a white patch on the throat. The rest of the under parts are yellowish buff, barred with black. The feathers extend to the tips of the toes. These birds breed very early in the year. At Princeton, New Jersey, I took a nest with three nearly fresh eggs, on February 19th, 1879. This was high up in an old chestnut, where a large limb, broken by the wind, was hollowed by decay. This had evidently been occupied by a gray squirrel. It was six inches deep and about a foot in diameter. The white eggs were nearly two inches and a quarter long and one inch and four fifths in their other diameter. The nests are more frequently made in deserted crows' and hawks' nests.

The birds occupy that part of Eastern North America east of the Mississippi Valley, from Labrador south, and are resident, breeding throughout their range. While beneficial in destroying many of the smaller rodents as well as reptiles, these birds are very destructive to poultry and many wild birds; quail, doves, even the larger hawks as well as song birds forming no inconsiderable part of their diet.

This geographical race of the Great Horned Owl occupies Western North

GREAT HORNED OWL.

America from Mexico north to Manitoba and British Columbia, and has been
taken in Illinois and Wisconsin.

Western Horned Owl.
Bubo virginianus subarcticus (Hoy).

It is much like its ally in general appearance, but
lighter, the yellows and buffs of that bird being replaced
by grays and white.

This is another race of the Great Horned Owl, occupying Labrador
and the Hudson's Bay country and tracts to Alaska. It
is a much darker bird in general tone, the black and
dusky colors prevailing over the browns and buffs.

Dusky Horned Owl.
Bubo virginianus saturatus
Ridgw.

The Saw-whet Owl is the smallest of the owls found in the region
treated of, being about eight inches long. Its size and round head without
ears should suffice to identify the bird.

Saw-whet Owl.
Nyctala acadica (Gmel.).

The adults are a peculiar shade of drab above, freely
streaked on the forehead and on the margins of the face
discs with white. The sides of the head and back of the neck are more
broadly streaked with white. The shoulders are broadly marked with white.
The tail has two or three narrow broken bars of white, and some of the
larger wing feathers are similarly marked. The lower parts are white broadly
streaked with drab of a warmer shade than that of the upper parts. The
feathers below the tail are white. The feet are feathered to the toes and the
eyes are yellow, surrounded by well defined facial discs of obscure white and
drab.

Young birds are plain brown above, rather darker than the adults. The
facial discs are dusky with white or buffy edgings. The breast is like the
back and the belly and feathers below the tail are yellowish buff. The birds
breed in deserted woodpeckers' holes and natural hollows in trees, and at
times in abandoned squirrels' nests. From three to five pure white eggs are
laid. They are about an inch and a fifth long and one inch in their other
diameter.

This is a migratory species breeding from Northern New York and New
England northward through the British Provinces. In winter it is found as
far south as Virginia. It is a strictly nocturnal bird, rarely moving about by
day, but passing that period asleep in some dark hemlock or cedar grove.
So profoundly does it sleep, with its bill and face buried in the feathers of

SAW-WHET OWL. ADULT.

the back and shoulders, that in a cedar grove, near Princeton, New Jersey, in December, 1878, I caught ten of the birds alive. This was of course unusual, but almost every winter I meet with some of these little owls in similar localities. Last winter, 1896 and 1897, I saw but four.

This is a more boreal bird than the Saw-whet Owl. It is very similar to that bird in general appearance, in color and markings, and the immature birds have a first plumage closely corresponding to the young of the Saw-whet. But the birds are so much larger, averaging ten inches and a half in length, as not to be confounded with their much smaller congener. The eyes are light yellow.

Richardson's Owl.
Nyctala tengmalmi richardsoni (Bonap.).

But little is known in regard to their breeding habits but presumably they are not unlike those of the Saw-whet. The eggs are larger but the same color and shape.

The birds are found throughout Northern North America and are the prototypes and close relatives of a similar Owl, Tengmalm's, *Nyctala tengmalmi*, that inhabits the boreal parts of the Eastern Hemisphere. Richardson's Owl is found in winter as far south as the northern border of the United States.

The Great Gray Owl is the largest of the dark colored owls found in our region. It is about twenty-seven or eight inches in length and without "horns" or "ear tufts." The eyes and bill are yellow and the feet heavily feathered to the tips of the toes.

Great Gray Owl.
Scotiaptex cinerea (Gmel.).

The general color is dusky with a fine mottling of white giving an effect of transverse barring. The darker color prevails above and the lighter color below and on the very prominent facial discs; where the barring takes the form of more or less defined concentric rings.

The call notes of these birds are said to add to the solemn dignity of the primeval woods of the northern solitudes.

The birds breed in trees but we have no adequate description of their nests. The eggs are two or three in number and white in color. They are about two inches and an eighth long, and nearly an inch and three quarters in width. They are found in the Hudson's Bay region and northward to the tree limit during the breeding season, migrating nomadically to the northern border of the United States during the colder portions of the year.

The Barred Owl is perhaps the most frequently seen and heard of the larger owls, and in certain localities is an abundant bird. It is a large bird,

Barred Owl.
Syrnium nebulosum(Forst.).
generally more than twenty inches long, has a round head, prominent and well defined facial discs, surrounding *dark-brown eyes* having *black pupils.* In general appearance it is a magnified Saw-whet. Its prevailing color above is grayish brown, the feathers being marked with two or more white or buffy white bars on each one. The facial disc is gray with concentric rings of dusky brown.

The breast is barred much like the upper parts. The rest of the under parts are grayish or buffy white striped broadly with dusky brown. The feet are feathered to the tips of the toes. The tail has narrow bars of white, seven or eight in number. The larger feathers of the wings are spotted with white.

The birds usually breed in hollows in trees but occasionally an old nest of some large bird, hawk or crow, is used. From two to four white eggs are laid. They are nearly two inches long and over an inch and three fifths in their smaller diameter.

The birds are found from north of Florida and the Gulf States, throughout Eastern North America to Nova Scotia and Manitoba. They are resident except at the more northern parts indicated.

In contradistinction to the deliberate *who who whowho* of the Great Horned Owl, this bird has a variety of notes and is frequently heard during the daytime, in the wilder parts of the country. These cries are given with little or no interval and are emphatic. They are varied by sounds that can only be likened to a demoniacal laugh, that is rung through many changes. In their shrillness and high pitch these sounds have a singularly weird and uncanny effect heard on moonlight nights in some lonely camp.

The birds are great hunters and destroyers of the smaller rodents, and frogs and lizards figure in their diet. The destruction of such petty devastators amply repays and much more than offsets the occasional fowl or small bird killed by these owls.

The Florida Barred Owl is a geographical race of the bird just described
Florida Barred Owl.
Syrnium nebulosum alleni
Ridgw.
and is similar in appearance, but smaller and darker colored, and *has the toes but partially feathered* or *nearly naked.*

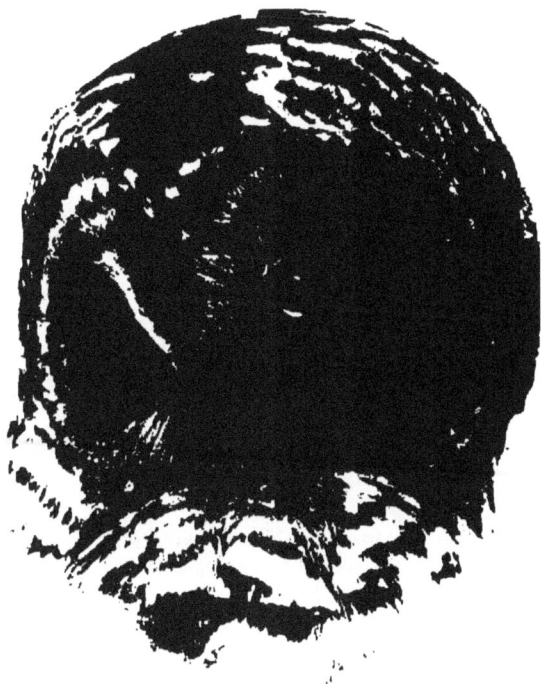

BARRED OWL.

The general economy and habits seem identical. It is found in Florida and on the coast of the South Atlantic and Gulf States to Texas.

A medium sized owl about fifteen inches in length, with conspicuous *ear tufts* or *horns, rather near together*, of a general gray color with a strong ad-

American Long-eared Owl.

Asio wilsonianus (Less.).

mixture of buff and with *yellow eyes*. Such is the Long-eared Owl.

The upper parts are dusky brown finely mottled with white, the bases of the feathers being buff. The facial disc is buffy behind the eye and gray in front with a strong admixture of black. The forehead is dusky finely mottled with white. The ear tufts are dusky brown bordered with buff or gray. The lower parts are gray and grayish buff striped on the breast with dusky. The sides and belly are both striped and barred in an irregular manner with dusky brown, each feather having a dusky line each side of its shaft and several more or less defined cross bars proceeding from it at right angles.

The birds generally use some abandoned nest of hawk, crow, or squirrel. Here from three to six white eggs are laid, about an inch and three fifths long and over an inch and a quarter in their other diameter.

The birds are found throughout North America. They range as far north as Nova Scotia and Manitoba, and breed thence southward to the Gulf States.

They are mice hunters and are nocturnal in their habits, usually spending the day in dark evergreen woods or cedar swamps.

The Pigeon Hawk is a true falcon of diminutive proportions, being about twelve inches long, a little larger and heavier than the Sparrow Hawk. This

Pigeon Hawk.

Falco columbarius Linn.

bird is really more of a sparrow and small bird hunter than the bird that bears the name Sparrow Hawk, birds to the size of a Crow blackbird forming the main part of the Pigeon Hawk's diet. A few mice and other small rodents, as well as some insects, are also eaten by these birds.

The Pigeon Hawk, in the adult plumage, is grayish blue above, having a more or less distinct brownish red collar or band on the neck. The tail has a white terminal and three or more well defined grayish white bars. The larger wing feathers are barred with white. The under parts vary in differ-

LONG-EARED OWL.

ent individuals from cream color to deep yellowish buff, and are streaked with dusky or black, except on the throat.

In immature birds the under parts are much the same in appearance. The upper parts however are dusky brown, and the broken band on the back of the neck is buff in color, as is the barring to the wings and tail. The terminal bar of the tail is whitish.

The nest is placed in trees in hollows in the limbs, and on ledges on the face of cliffs. The eggs, four to five in number, are about one inch and three fifths long by nearly an inch and a quarter in their other diameter. They vary much in color and markings, some being almost white, marked with reddish brown and chocolate, others being deep reddish brown, marked with darker shades of brown and chocolate. There is every possible gradation, between these two extremes. The birds are found throughout North America. They breed from the Northern United States northward, and winter from the South Atlantic and Gulf States to Northern South America.

This hawk resembles in its habits the Broad-winged Hawk, but is much more shy, and is a little larger in size, being about seventeen and a half inches in length. It is a woodland bird, those which I **Short-tailed Hawk.** have personally seen frequenting " hammocks "and heavy Buteo brachyurus Vieill. cypress growths. In this country it has so far been found only in Florida, though it ranges south through parts of Mexico into South America. In Florida it is of regular occurrence, though not common.

There are two distinct color phases of this bird, and it is not improbable that they may be found to correlate with the sexes on further careful study.

In one of these, that which seems to prevail largely among male birds, the entire bird is very dark in color. The upper parts being darkest, almost black, have a slight lustre, the forehead is indistinctly white, and the tail above is a little lighter than the back, and barred with the color of the back. The lower surface of the tail and wings is silvery gray. The lower parts are like the back, but not quite so dark.

The lighter phase, which so far as I have examined the birds seems to prevail among females, is characterized by the upper parts being brown, with a strong suffusion of gray or slate color. The tail is barred with black and tipped with white, and its under surface is silver gray. The forehead is obscurely whitish and the lower parts are white with chestnut brownish markings on the sides of the breast. In both phases the eyes are brown.

BROAD-WINGED HAWK.

We have no very adequate account of the breeding of these birds. Mr. Charles J. Pennock found a pair nesting in trees near St. Mark's, in Florida, early in April. They had laid but one egg, which he secured. It is about two inches and three twentieths long and an inch and three fifths in its smaller diameter. It is bluish white in color, speckled and marked on the larger end with reddish brown for about one fourth of its surface. A few finer spots are irregularly dispersed over the rest of the egg to its smaller end.

A nest which I found near Tarpon Springs was in the edge of a hammock in a gum tree about forty feet from the ground. This nest was not completed, as the birds were procured for study, and are now in the Museum of Comparative Zoölogy, at Cambridge, Massachusetts.

The Broad-winged Hawk is our smallest representative of the group of the heavy stout hawks known as the Buteos. The males are about fifteen inches long, and the females often exceed seventeen inches in length.

Broad-winged Hawk.

Buteo latissimus' Wils.).

The upper parts are dusky brown, with grayish suffusions and lighter edges to the feathers. This becomes more apparent on the head and sides of the face, which are streaked with reddish brown.

The tail above is almost black, and has the *tip* and *two broad bars* grayish white. The outer tail feathers have silvery gray under surfaces. The under parts are white, streaked on the throat and heavily barred on the breast and chest so as to almost obscure the ground color, with deep brown, or yellowish buffy brown, varying in individuals. This color becomes more broken on the lower chest, which is distinctly barred ; farther back the white begins to prevail over the brown till the belly and feathers below the tail are immaculate.

Immature birds are much like the adults above, but the tail is grayish brown, and has five obscure darker bars and a narrow white tip. The lower parts are white or light buff *streaked* with brownish black.

These birds breed throughout the region under consideration from New Brunswick to Texas, extending their migrations to the West Indies and Northern South America. They nest in trees from twenty to fifty feet from the ground, and lay from two to four eggs. These are about an inch and nine tenths long, and nearly one inch and three fifths in their other diameter. They vary much in appearance, being white or buff in ground color and marked with spots of varying degrees, and shades of brown.

In the Woods.

This is essentially a woodland hawk, and feeds mainly on small animals, such as mice, frogs, and insects, rarely preying upon birds. It is one of the commoner hawks, and in the fall migrations, when they become gregarious, sometimes enormous flocks may be seen, passing southward, high in the air.

This is a bird of both woodland and open, and while not perhaps as much of a field hunter as the Red-tailed Hawk, yet is frequently seen on the wooded borders of meadows. He is one of the so-called "Chicken Hawks," but is really no enemy to the farmer, as careful investigation has shown that small animals such as field mice, shrews, frogs, etc., form by far the larger part of his diet. He is not a bird or chicken hawk, but a mouse hunter.

Red-shouldered Hawk.
Buteo lineatus (Gmel.).

You will know him in any plumage by his reddish shoulders. Sometimes in immature birds this marking may be somewhat obscured, and it is always duller in these than in birds in full plumage, but *it is always present*. The adult birds are very striking, being characterized by black or dusky wing and tail feathers, the former with many, the latter with four or five white bars, and a white tip. The lower parts except the throat, which is streaked with dusky, vary from very bright to dull burnt sienna, barred with narrow stripes of pure white. The upper parts are dusky brownish suffused with gray, and each feather is more or less edged with reddish brown or buff, parts of the shoulders being *bright* reddish brown, as described before.

In immature birds the upper parts are much the same, but the wings and tail are dark grayish brown, both with obscure bars of dusky black. The red on the shoulders is not so bright, and it is less extensive. The under parts are white or grayish streaked with dusky.

They breed in trees from twenty-five to seventy-five feet from the ground, and lay from three to six dull white eggs marked more or less profusely with specks and blotches of reddish brown and umber. These are two inches and an eighth long, and about an inch and seven tenths broad.

The birds are found throughout Eastern North America as far north as Manitoba and Nova Scotia, and south to the Gulf States. They are resident through most of this area, and breed throughout their range.

The Florida Red-shouldered Hawk is the geographical form or race, taking the place of the Red-shouldered Hawk on the Atlantic coast, as far north as South Carolina, in the whole of Florida, and along the Gulf coast to Eastern Texas. It is a bird of smaller size. The adult plumage is characterized by its **Florida Red-shoul-** lighter color, by a general absence of bright rufous, both **dered Hawk.** above, on the head, and below. These parts are all suf- Buteo lineatus alleni Ridgw. fused with a grayish or whitish tinge. This is especially marked on the head, sides of the face, and on the lower parts. The reddish on the shoulders and lower parts is dull and often becomes buffy. Immature birds are smaller and generally darker than immature representatives of the more northern race.

The Goshawk, one of the birds of falconry fame, seems typical of the group about to be discussed, and a word as to its life and methods will apply **American Gos-** to all three of the Accipiters that are found with us. Long **hawk.** slim birds with rather short and rounded but powerful Accipiter atricapillus wings, they steal upon their prey, flying rather low or (Wils.). near the ground. They hunt by preference in the thick woods, but are frequently seen in the open. They are fearless and daring and do not seem to regard man when in pursuit of their bird victims. Anyone who has shot much has now and again seen one of these bold fellows steal a bird that had just been killed by the sportsman. Before the report had fairly ceased to echo and while the smoke still curled from the muzzle of the gun, seizing a bird nearly as large as himself the fearless robber carries him away almost before you are aware of his presence. These are the hawks whose depredations spread panic among the poultry and are a constant source of dread to the woodland birds, whose clamors often announce the advent of one of them ere you in your walk have discovered it.

Of these three birds, the Sharp-shinned Hawk, Cooper's Hawk, and the Goshawk, the last is much the largest, being about two feet long, the female somewhat bigger.

The adult birds have the top of the head clear black, the bases of the feathers being pure white. There is a white line over the eye extending back. The rest of the upper parts are bluish gray, each feather with a blackish streak along its shaft. The tail is like the back in color becoming more dusky on its outer feathers, and crossed by four or more obscure dusky bars, and with whitish tips to the feathers. The lower parts are white, irregularly barred with slaty gray, and the feathers of the throat and breast have dusky streaks along their shafts. The eyes are orange yellow.

In the Woods.

Immature birds are dusky grayish brown above, the feathers margined with rufous; those on the head are edged or streaked with a similar shade. The tail and wings are brownish gray, darker than the back, the tail being distinctly barred with black. The lower parts are white or cream color streaked with dusky brown or black. The eyes are yellow. The birds nest in trees. They lay from two to five nearly white eggs, sometimes faintly marked or splashed with pale brown. They are about two inches and a third long and one inch and three quarters in width. The birds are northern in their distribution, breeding from the Northern United States northward, and wintering regularly as far south as the Middle States.

The Sharp-shinned Hawk is the smallest of the trio of our Accipiters. The males are noticeably smaller than the females, averaging about eleven

Sharp-shinned Hawk.
Accipiter velox (Wils.).

inches and a half in length, whereas the females are fully two inches longer. Adult birds have the upper parts bluish slate color, the throat white with dusky streaks, and the rest of the lower parts white barred with pale reddish brown or buff. The *tail* is almost *square* with dusky barring and white tip, and the larger wing quills are barred with dusky brown or black. The eyes are yellow or orange yellow.

Immature birds are dusky brown above, with reddish brown edgings to the feathers, and the tail and larger feathers of the wings much the same as in the adult birds. The lower parts are white streaked more or less profusely with dusky or dark reddish brown. The eyes are pale yellow.

Cooper's Hawk at first glance seems an enlarged copy of Sharp-shinned Hawk, and differs but little from that bird save in size; the males are about

Cooper's Hawk.
Accipiter cooperii (Bonap.).

fifteen inches and a half in length, and the females are more than three inches longer. Adult birds have a dusky crown. The rest of the coloring and markings are about the same as in the Sharp-shinned Hawks. Immature birds resemble that phase in the smaller bird. The tail is rounded, the outer feathers being much shorter than the next and so on, the middle ones being the longest. The eyes are yellow.

The two birds in any plumage may always be distinguished by the shape of the end of the tail; in the Sharp-shinned this is square, and in the Cooper's Hawk it is noticeably rounded.

Both birds breed rather high in trees, laying from three to six eggs. Those of the smaller bird are about an inch and a half long by an inch and a fifth in their other diameter. They vary in color from bluish white to cream, and are distinctly spotted or washed with varying shades of brown. Cooper's Hawk's eggs are bluish white in color, sometimes immaculate, but generally spotted with pale brown. They are about an inch and nine tenths long and rather more than an inch and a half in their smaller diameter. Both birds are found throughout North America, breeding in the United States and northward.

Cooper's Hawk is perhaps more northern in its summer distribution. Both winter from Central New England south to Mexico and Central America.

The introductory remarks in regard to the habits of the Goshawk are applicable to both of its smaller relatives.

A much enlarged Barn Swallow, a close copy of that bird in shape and even in action ; a bird two feet long, with all the graceful lines of the Swal-
Swallow-tailed low, from the forked tail and pointed wings to the short
Kite. neck ; such, in a general way is the first impression one
Elanoides forficatus (Linn.). gets of the Swallow-tailed Kite.

In color the birds are as noticeable as in form. Adults have the head, neck, and a band across the rump, and the under parts pure white. The back, wings, and tail are lustrous black, with some iridescent areas. The smaller inner quill feathers of the wings are white, with blackish tips.

These birds build nests in high trees, where they lay from two to four eggs. These are white or cream colored, variously spotted and marked with different shades of brown. The eggs are more than an inch and four fifths long, and nearly an inch and a half in width.

The birds are more frequent in the interior than elsewhere in Eastern North America, ranging as far north as Minnesota. On the coast they range commonly to the Carolinas, and casually from there north to Southern New England. They winter in Central and South America. They breed irregularly and locally throughout their North American range.

Twenty-five years ago it was possible almost anywhere in our Eastern States where there was woodland to see some Wild Pigeons every spring and fall. In my early studies of birds in 1870, and before, there were Wild Pigeons about Cambridge in Massachusetts. Close to the town, in the vicinity of Mount Auburn, a few bred every year. Later in 1875, in the vicinity of Princeton, New Jersey, there were many Wild Pigeons passing in flocks, especially in the fall. These were fragments of the great hosts of these birds, which once formed such a feature of the bird fauna of Eastern North America. Most of these small bands are but memories in the minds of the students and lovers of birds in those days. Personally I have seen but eight Wild Pigeons in the last fifteen years, though constantly working in fields where they were once abundant. They still exist in what would to us seem large flocks in Michigan and some other points in the West and North. But it seems the consensus of opinion that their extermination is almost completed and that with the bison and paroquet they will soon be relegated to the position of exterminated elements in our fauna. Those who care for accounts of their wonderful migrations and breeding grounds early in this century, must consult Audubon and Wilson.

Passenger Pigeon.
Ectopistes migratorius (Linn.).

The adult Wild Pigeon is about sixteen inches and a half long, and in appearance somewhat like a Mourning Dove. The upper parts are slaty blue, having iridescent metallic areas on the back and sides of the neck, The central pair of tail feathers are dusky and the others are dark at their bases, shading into an area of slaty blue, which in its turn shades into the pure white of the tips. The chest and breast are reddish fawn color, changing into a more pinkish shade on the sides. The belly and feathers below the tail are white. The female resembles the male but has less iridescent sheen, and is duller colored generally. Immature birds lack the iridescence and are still more subdued in tone, each feather of the breast and back being tipped with grayish white and those of the wings edged with reddish brown.

The nests are rude flat structures of twigs and sticks placed in branches of trees. One or two white eggs are laid ; they are nearly an inch and a half long and somewhat over an inch in their smaller diameter.

The Wild Pigeon was formerly common throughout Eastern North America, as far north as Hudson's Bay. It bred at various points, generally in great communities, and wintered in the more southern part of the range.

The Quail-Doves are a group of tropical pigeons that live in the dark woods on or near the ground. They are silent and stealthy in their habits and even when comparatively common are rather difficult to observe. Three species of which this one seems to be the only regular visitor have been recorded from the Florida Keys. It is a bird about eleven inches long, bright chestnut above, with the back and neck having green and purple metallic lustre. The breast is dull chestnut and the belly white.

Key West Quail-Dove.
Geotrygon chrysia Bonap.

The nest is built in low trees or bushes where two white eggs are laid. They are less than an inch and a quarter in length and rather more than nine tenths of an inch in their other diameter.

This is a cinnamon colored Dove, darker on the back, where there are purplish reflections. It is about ten inches long, and is a native of Tropical America and the West Indies, and has been recorded once from Key West.

Ruddy Quail-Dove.
Geotrygon montana (Linn.).

The Blue-headed Quail-Dove is a Cuban species of irregular occurrence on some of the Florida Keys. It is olive brown above and dull rufous brown below. The top of the head is dull blue bordered by black. There is a conspicuous white stripe below the eye, starting from the throat and extending to the back of the head.

Blue-headed Quail-Dove.
Starnœnas cyanocephala Linn.).

The English Pheasant was introduced into Europe and the British Islands from Western Asia, at so remote a period that the exact date is largely a matter of conjecture. On the British Islands these birds were held in great estimation before the time of the Norman Conquest.

English Pheasant.
Phasianus colchicus Linn.

At various times during the past half century it has been imported from England and Europe to this country and bred in captivity. During the past ten years it has been liberated at various points in the Eastern United States and must now be regarded as an element in the bird fauna of that region. In my rambles through the woods in both New York and New Jersey I have on several occasions met individuals of the species that were apparently wild or naturalized birds.

The birds are frequently exposed for sale in the markets of our larger cities and are too well known in appearance to need further notice here.

It seems probable that the domestic Turkey is a descendant of birds taken at the time of the conquest from Mexico to Europe, and brought from thence by the early settlers to the Eastern United States. These birds were crossed by the natural wild stock, then abundant in the country as far north as Maine.

Wild Turkey.
Meleagris gallopavo Linn.

The Wild Turkey is now restricted to the regions from Pennsylvania south to Florida, having become exterminated in its former northern haunts and now becoming yearly more uncommon where it still occurs.

The adult male birds are about four feet long and may be known from the familiar domestic bird of dark color, by the reddish brown or chestnut tips of the feathers of the rump and tail, which are generally white or cream color in the domestic bird. *Also by the wattles of the head* which are much less developed in the wild than in domesticated birds.

The female birds are much smaller than the males but the same features will serve to distinguish them.

The nest of the Turkey is placed on the ground under a bush or at the base of a tree. The eggs vary much in number, from six to twelve and sometimes more. They are pale buff thickly speckled with fine brown dots, and are about two inches and two fifths long and an inch and nine tenths in their other diameter. The food of Wild Turkeys consists of seeds, nuts, grain, insects, and the smaller reptiles and little fish.

This is the geographical race of Wild Turkey occupying Southern Florida and ranging as far north at least as Sumpter County. It is a smaller and more intensely colored bird than the Wild Turkey of the more northern regions, and has *broken white markings* taking the place of the white bars, conspicuous on the larger wing feathers of the Wild Turkey.

Florida Wild Turkey.
Meleagris gallopavo osceola Scott.

Of our native game birds the Ruffed Grouse, Partridge, or Pheasant, as the bird is known in its more southern range, is a bird eminently calculated to maintain itself against the constant onslaughts of sportsmen. Its whole method of life, the localities it frequents, and its extreme wariness where much hunted, seem to insure its continued existence where other game birds have become scarce or have disappeared. Even in the vicinity of large towns and cities, where there

Ruffed Grouse.
Bonasa umbellus (Linn.).

is any considerable area of woodland, it is still comparatively common and in the rough mountain regions and the large forests of the North and West it is abundant. Its food supply, during the warmer parts of the year, consists of insects and the smaller wild fruits, and in the rigor of winter, buds, leaves, and seeds are the staple diet.

A large bird a foot and a half in length, the male is of general rusty brown color above. The feathers of the top of the head are somewhat elongated, forming, when erected, a prominent crest. On each side of the neck is a noticeable tuft of black feathers. The tail varies in color from grayish to rusty brown. There is a broad black band near the end and the tips of the feathers are gray. The rest of the tail is barred and mottled with black in an irregular way. There is a dusky band on the breast and the throat is brownish buff. Back of the breast band the lower parts are white, or grayish white, with buffy suffusions, and barred, obscurely on the breast and belly and clearly on the sides with dusky brown or blackish.

The female is smaller, and the neck tufts are much less prominent, sometimes being obsolete.

The nest is placed on the ground close to a stump or fallen log. From six to ten eggs are the usual number laid. They are plain buff in color and are about an inch and a half in their large, and an inch and a fifth in their small diameter. These birds range from Vermont to Virginia and on the higher Alleghanies to Georgia and Tennessee. They extend westward to the Great Plains.

Canadian Ruffed Grouse.
Bonasa umbellus togata (Linn.).
     This is the geographical race of the Ruffed Grouse that is found in the northern half of the Northern New England States, Northern New York, and thence north to Hudson's Bay. The birds are gray in tone above instead of reddish brown, and are much more distinctly barred beneath.

The Spruce Partridge, or Canada Grouse, is a smaller bird than the Ruffed Grouse, being about fifteen inches and a half long.

Canada Grouse.
Dendragapus canadensis (Linn.).
     The general effect of the male is black and white. The upper parts are barred with black and grayish. The tail is black tipped with reddish brown. There is an exposed bare area above the eye, which is bright red in life. The under parts

207   RUFFED GROUSE. ADULT MALE.

are glossy black, with varied white markings. The sides are mottled with black and grayish white. There is a broken band of black and white across the upper breast, defining that area from the throat.

The female is browner than the male in general appearance. There are no defined black areas on the under parts, but the throat is barred with pale reddish brown and black, and this gradually gives way on the breast to dusky or black broken by the grayish tips of the feathers. The upper parts are similar in pattern to those of the male, but everywhere browner. The tail is much the same as in the male, but is mottled with reddish brown and tipped with the same color more narrowly. The nest is placed on the ground in the woods much as is that of the Ruffed Grouse, and the eggs are buffy, specked with brown. They are an inch and seven tenths long and an inch and a quarter in their other diameter. These birds are found east of the Rocky Mountains in Northern North America, from Northern New England and Minnesota northward.

ACROSS THE FIELDS.

ACROSS THE FIELDS.

THE Stone-Chat or Wheatear is a European bird that is represented in the fauna we have under consideration, in Labrador and in the Province of Quebec, Canada. It has been recorded as an accidental straggler in Nova Scotia, Maine, Long Island, and New Orleans Louisiana. The birds are about six inches and a quarter long. Their prevailing colors are black, white, and gray, in sharply defined areas. The upper parts are light grayish, becoming white on the forehead, rump, and the region above the roots of the tail. There is also a white stripe above the eye. The sides of the face and the wings are black, as is the terminal third of the tail. The rest of the tail and the under parts are white, the latter suffused on the breast with buff. The female is similar but the colors are more obscure, the black areas being dusky and the white parts suffused with buff.

Wheatear.
Saxicola œnanthe (Linn.).

In winter both the adult and immature birds are plain brown above, except the rump and basal portion of the tail, which are white. The wing feathers are dusky in the male and grayish in the female, edged with light reddish brown. The black of the tail is replaced by dusky and tipped with light buff. The lower parts are obscure brownish buff, darkest on the breast and chest.

The nest is placed on the ground among rocks and stones ; it is built of grasses, moss, and other plant fibres, and lined with finer material and feathers.

The eggs are pale greenish blue, unmarked, and vary from three to seven in number. They are more than four fifths of an inch long, and about three fifths of an inch in their other diameter.

The Brown Thrasher is of all our smaller birds perhaps the most noticeable. Frequenting the edges of the woodland, open fields, and hedgerows, his alert movements, brilliant vocal power, and reddish brown coat alike attract the eye and ear.

He is a large bird, too, more than eleven inches long and is almost as inquisitive as his cousin the Catbird.

Brown Thrasher.
Harporhynchus rufus (Linn.).
When seen in some thicket you are at once reminded of a wren ; the same characteristic jerky, quick, and nervous movements are displayed.   But when the bird, mounting to some high branch, abandons himself to song, the performance is so varied and brilliant that all else is for the time forgotten.

The birds are bright cinnamon brown above, with some white markings on the wings forming two more or less distinct bars.   The under parts are white streaked with dusky brown on the breast, sides, and flanks.   The eyes of the Brown Thrasher are yellow, or orange.   In the fall the lower parts and the white bars of the wings are much suffused with buff.

The nest is generally placed on or near the ground, but sometimes at a considerable height in bushes and in the lower branches of trees.   It is built of coarse twigs and grasses, lined with fine rootlets and plant fibres.   The eggs vary from three to five in number.   The ground color varies from greenish white to pale buff.   The darker markings on the ground color are minute and close together, giving the eggs a general reddish brown appearance.   They are somewhat more than an inch long and nearly four fifths of an inch broad.

The birds are found throughout the area under consideration as far north as New Brunswick.   They breed from the Gulf States north, and winter from Virginia southward.

The Pipits, or Titlarks, are small birds, about six inches and a half long, of a general yellowish brown appearance.

American Pipit.
Anthus pensilvanicus (Lath.).
They breed far north, and when seen in the Eastern United States are migrating, or wintering, and are then gregarious.   You may find them in meadows or fields, but they seem to prefer newly turned ground, or the districts of sparse grass near the seacoast.

The upper parts are dull brown, with a grayish tinge.   There is a stripe of light buff, above the eye.   The feathers on the top of the head and back have distinctly darker centres.   The wings and tail are darker, their feathers are edged with buffy gray.   There are two wing bars of a similar shade.   The terminal half of the outer tail feather is grayish white.

The lower parts are light warm buff, palest on the upper throat.   The

BROWN THRASHER.

NEST AND EGGS OF BROWN THRASHER.

breast is generally distinctly streaked with dusky brown, and the sides and flanks are often more obscurely marked in the same way.

AMERICAN PIPIT.

The nest is placed on the ground, is made of mosses and plant fibres. The eggs, four or five in number, are bluish white, evenly and profusely dotted with reddish brown. They are about four fifths of an inch long, and less than three fifths of an inch wide.

The Meadow Pipit of Europe is sometimes found in Southern Greenland. It is a smaller bird than the American Pipit, about six inches long. In general color it is not unlike that bird, but is darker and **Meadow Pipit.** more olive brown above and *distinctly streaked with black.* Anthus pratensis (Linn.). The lower parts are buffy white, and the breast, sides, and flanks are streaked with dusky brown. Its breeding habits are very similar to those of the American Pipit.

Sprague's Pipit is a bird about six inches and a half long. It is streaked above with dusky and grayish buff. There is much white on the outer tail feathers. Below it is light buff or white tinged with buff, **Sprague's Pipit.** which is clearest on the breast, and streaked in that region Anthus spragueii (Aud.). with dusky brown.

It is a bird of the interior plain region of North America, breeding from

Central Dakota northward. It has been recorded as accidental from near Charleston, South Carolina.

This bird is another European species that has been recorded from Southern Greenland. It is a bird about seven inches long with a rather long tail and of slim build. It has a black crown extending well down on the back of the neck. The forehead is white and there is a white line above the eye and the sides of the head are white. The back and shoulders are grayish. The tail is black with some of the outer feathers white. The throat and breast are black and the rest of the lower parts white.

White Wagtail.
Motacilla alba Linn.

It breeds on the ground generally near water. The nest is made of dry grasses and like material. The eggs are nearly white, profusely speckled with brown. They are nearly four fifths of an inch long and about three fifths of an inch in their other diameter.

The Yellow-breasted Chat is a bird of such marked characteristics as to have achieved fame. To say that these birds are eccentric and totally unlike any of their kindred conveys but little to such as have not met them. With all the curiosity of the Catbird, these birds seem ever alive to the fact, that they too attract much attention. And while you may hear many Chats in the regions they frequent, as they are very loquacious, your opportunities of seeing them will be in a large inverse ratio. Very wary of exhibiting themselves to the watcher, they are constantly near, and their medley of curious notes rings out so near at hand that you feel sure of seeing the singer. However, persistent observation is sure to be rewarded and sooner or later the song, that has so often tantalized you from out some tangle of bushes, will be heard overhead and close to you. Looking, you will see a bird in the air, apparently careless of the onlooker, engrossed in a performance at once wonderful and grotesque.

Yellow-breasted Chat.
Icteria virens (Linn.).

With dangling legs and slowly flapping wings, with every feather seemingly awry, and with an uncertainty of flight strangely like some large moth or butterfly,—such a mixture of curious notes is poured out as has no kind of parallel in our bird acquaintance. This is no soft melody that one has to be near to hear, but a series of loud jerky, detached notes, now whistles, now clucks, and again croaks and chuckles, that defy imitation, musical or otherwise.

YELLOW-BREASTED CHATS.

NEST AND EGGS OF YELLOW-BREASTED CHAT.

When you do see him clearly you will find a bird about seven inches and a half long. All the upper parts and the wings and tail are green with an

YELLOW-BREASTED CHAT.
ABOUT TEN DAYS OLD.

olive tinge. On the sides of the head this color is interrupted by a clear white line extending from the nostril above and back of the eye. The region just in front of the eye is almost black. There is a white ring about the eye and a more or less distinct white line bordering the olive green sides of the head and face, and defining those regions from the throat. This region and the breast and chest are bright yellow, extending in some individuals even farther back. The belly and feathers below the tail are white and the flanks are grayish olive.

The Chats frequent dense tangles of undergrowth in open clearings or fields. In such places they build their nests rather near to the ground. These are rather bulky structures of large dead leaves, strips of bark and

YELLOW-BREASTED CHATS.
ABOUT TWO WEEKS OLD.

220

grasses, well woven together and lined with finer material. Three to five eggs are laid. They are white, speckled all over, but not closely, with reddish brown. They are about nine tenths of an inch long and less than seven tenths of an inch in their median diameter.

YELLOW-BREASTED CHAT.

Chats are found from the Gulf States as far north as Massachusetts and Southern Minnesota. They breed throughout this area, and winter in Central America.

The Prairie Warbler is a bird about four inches and three quarters long. As its name indicates, it frequents rather open places, bushy fields, or clearings where a new growth is beginning.

Prairie Warbler. The adult male is clear olive green above. The back
Dendroica discolor (Vieill.). is spotted with reddish brown in a defined area. The wings and tail are dusky. There is a yellow wing bar on each wing. The inner webs of the outer tail feathers are white almost to the tips. There is a line of bright yellow above and another below the eye. The region in front

and behind the eye is black, and there is a broader black stripe from the corner of the mouth across the cheeks. The lower parts are bright yellow, the sides and flanks striped with clear black.

The female is similar but duller and the reddish brown of the back is more obscure and often wanting. The black striping of the male is replaced by dusky or grayish markings. Immature birds resemble the adult female but are strongly tinged with ashy. The bar of the wing is indistinct or often altogether lacking. The nest is built in tangled bushes or young cedars in rather open ground. It is composed of various vegetable fibres and plant down lined with finer material and long hair. Four or five eggs are laid. They are white, spotted and speckled, mainly at the larger end, with varying shades of brown. The eggs are rather more than three fifths of an inch long and less than half an inch in their smaller diameter.

The birds are found during the warmer portions of the year as far north as Southern New England and Michigan, from which points they breed locally as far south as Florida. They winter in Florida and the West Indies.

This is another Warbler that seems to choose the vicinity of cleared lands for its summer home. The birds are about five inches long.

The adult male has a greenish yellow crown, which is bordered with black on the sides. There is a broad area of black in front of the eye extending **Chestnut-sided** downward and defining the throat. The remainder of the **Warbler.** face and the sides of the neck are white. The back of the Dendroica pensylvanica neck is streaked black and gray, which becomes black (Linn.). striped with greenish yellow on the back. The wings and tail are dusky. The former with two greenish yellow bars and many of the feathers are edged with a similar color. The outer tail feathers have their inner webs white almost to the tips. The lower parts are pure white, the sides being *broadly and definitely deep rich chestnut.*

Female birds are similar in pattern but duller in color.

Immature and autumnal birds are light olive green above with wings and tail similar to breeding birds. The back is often obscurely streaked with dusky. The sides of the face are grayish, as is a ring about the eye. This same color prevails on the sides of the neck. The throat, breast, and belly are white shading into the gray of the sides, neck, and face. Many individuals show some trace of the deep chestnut of the sides.

The birds nest in low bushes. The nest is built of plant fibre and strips

PRAIRIE WARBLER.

CHESTNUT-SIDED WARBLER. ADULT MALE IN SPRING.

of bark, and lined with finer material and rootlets. The eggs are white with markings of varying shades of brown, mainly about the larger end. They vary from four to five in number, and are about two thirds of an inch long, by a little less than half an inch in their other diameter.

This bird is found as far north as Newfoundland and Manitoba. It breeds from Northern New Jersey and Illinois northward, and south on the Alleghanies to South Carolina. It winters in Central America.

Bell's Vireo is a bird of the interior of North America, where it is found from Texas to Minnesota extending eastward to Illinois, where it is common in localities. It winters as far south as Southern Mexico.

Bell's Vireo.
Vireo bellii Aud.

It is a small bird, being about four inches and three quarters long and similar in form to the White-eyed Vireo.

The general color of the upper parts is olive green shading into grayish olive on the back of the neck and becoming ashy gray on the crown. There are two wing bars. The region in front and a ring about the eye are grayish white. The under parts are white shaded on the breast, sides, and flanks with greenish yellow. The sexes are alike and immature birds are similar.

The nest is pensile or semi-pensile, built of bark and plant fibres and lined with fine grasses. It is placed generally in bushes but sometimes in the lower branches of trees. Four to six white eggs are laid which are sparsely speckled with dark brown and black. These are nearly seven tenths of an inch long and about half an inch in their other diameter.

The birds frequent rather open country, bushy fields, and thickets about the edges of swamps and along the banks of streams.

In the more Southern States where the Mockingbird is common, there is a bird of about the same size, shape, and general coloration. This is the

Loggerhead Shrike.
Lanius ludovicianus Linn.

Loggerhead Shrike or Butcherbird, a bird of very considerable vocal attainment, but with certain hawk like attributes. On some high perch, where a considerable territory can be carefully observed, he sits watching for his prey. This is generally a large grasshopper, a lizard, or small snake, and now and again when pressed by hunger some one of the smaller birds. Patience and alertness are his characteristics. Once detected he rarely fails to capture his

victim, which is speedily killed and generally quickly eaten.   But this Shrike,
be it with an eye to a future meal, or because surfeited and having, like more
civilized sportsmen, the destructive instinct so strong as to be unable to fore-

LOGGERHEAD SHRIKE. ADULT.

go an opportunity to slay, often seems to kill for killing's sake.   Then his
victim is not eaten, but is impaled on some sharp thorn or splinter, near the
place of capture.

The Loggerhead Shrike is about nine inches long ; a stoutly built bird
with a large head and a heavy hooked bill with a noticeable tooth on either
side.

The upper parts are gray and the wings and tail black.   The bases of
the larger wing feathers and the tips of the smaller ones are white.   The tail
feathers are generally tipped with white, and the outer one is almost white.
The region in front of the eye and narrowly across the forehead is black, as
is the region about the ears.

The lower parts are generally white, but not unfrequently grayish and
sometimes are obscurely and finely barred with dusky.   Immature birds have
more pronounced barring below, and obscure barring above, and are suffused
with brownish.

The birds build a substantial nest of twigs and strips of bark and other
plant fibres, lined with finer material.   These are generally placed in thorny
bushes or trees, rather low down.   An orange tree in the South or a locust
in the North is frequently chosen.   From three to five dull white eggs,
thickly marked and spotted, with varying shades of brown, are laid.   They

are a little less than an inch long, and rather more than seven tenths of an inch broad.

The Loggerhead Shrike is found in Eastern North America, west to the Plains. They breed from the Gulf States to Virginia and sometimes Southern New Jersey on the coast. In the interior they breed north to the Great Lakes and through Western Pennsylvania and Central New York to New Hampshire, Vermont, and Maine.

The Great Northern Shrike is almost a copy of the Loggerhead in color, but is so much larger, ten inches and a half long, as to be rarely confounded with that bird.

Northern Shrike.
Lanius borealis Vieill.
The salient color differences are a *whitish fore-head* and *the region in front of the eye is not so clearly black* as in the Loggerhead. The under parts are white or grayish white seldom immaculate, but generally barred with fine wavy lines of dusky or blackish.

GREAT NORTHERN SHRIKE. IMMATURE.

Immature birds have the upper as well as the lower parts obscurely and more heavily barred, and are suffused with brownish. The nest is built much in the same way as that of the Loggerhead, and the eggs are much like those of that bird in color and markings, but are appreciably larger.

The Great Northern Shrike seems more disposed to kill and eat small birds than does the Loggerhead, and the Sparrows and Snowbirds are

frequently harried by him.   He too appears to be unable to refrain from killing more than he can at once use, as witness his many impaled victims.   While
more frequently making sorties from his perch, he also stops in his flights as
he passes low over the ground to catch some bird or insect prey.   He too is
a song bird and perhaps of greater attainments than his relative.

The great Northern Shrike is of far northern distribution during the
breeding season and is only known as a winter visitor or migrant, in the
States from Virginia and Kansas northward.

The Rough-winged Swallow is about five inches and three quarters long.
The upper parts are brownish with a strong gray tone.   The *throat and*

Rough-winged   *breast* are nearly the same color as the *back* but of a some-
Swallow.   what lighter shade.   The belly is white.   Adult birds have
Stelgidopteryx serripennis   the *outer edge of the first wing feather serrate or rough*
(Aud.).   *to the touch.*   This is occasioned by the sharp and re-
curved webs of the feather, which forms minute, recurved hooks.

Immature birds do not possess this last characteristic but may be known
by their size and the color of the *throat* and *breast*, which is much like that of
the adults but generally has a distinct wash of rusty brown.

ROUGH-WINGED SWALLOW.

The birds nest in stone walls, under bridges, and in holes in banks, where
a loose structure of grasses and feathers serves as a nest.   Four to eight

white eggs are laid, which are nearly three quarters of an inch long and about half an inch broad.

These Swallows are found throughout the United States at large. In the area under consideration they are found as far north as Connecticut and Southern Minnesota, breeding locally from these points, southward into Mexico. They winter in the Tropics.

Bank Swallow.
Clivicola riparia (Linn.).

The Bank Swallow is a smaller bird than the Rough-winged Swallow, about five inches and a quarter in length. They are uniform dull grayish brown above, darkest on the head and wings and lightest on the rump. The *throat* and lower parts in general are *white.* There is a *broad band* in color like the upper parts *across the breast* and continuous with an area of the same color on the sides. Immature birds are much like the adults but have the feathers of the shoulders and the rump edged with buffy or grayish white.

BANK SWALLOW.

These birds are gregarious, breeding locally in suitable localities where sand banks afford the places adapted for their burrows. In such locations many birds form breeding colonies. The nest is a loose structure of grasses and feathers at the bottom of a burrow two or three feet deep excavated by the birds. The eggs vary from four to six in number and are white in color. They are nearly seven tenths of an inch long and a little less than half an inch broad. The Bank Swallow has much the same distribution in Eastern North America as the Rough-winged Swallow, but extends farther north. The birds are found throughout the Northern Hemisphere. They breed from the middle districts of the United States north to the tree limit. They winter as far south as Brazil.

Among the swallows of our region the White-bellied Swallow is noticeable. It is a field or marsh bird in its feeding habits for a large part of the time that it is with us, resorting to open woodlands and **Tree Swallow.** the vicinity of houses to nest and breed. In the woods Tachycineta bicolor (Vieill.). the deserted nests of woodpeckers and other holes and hollows in trees afford the birds nesting sites. About houses they often use, if allowed by the Wrens, Sparrows, and Bluebirds, the "bird boxes" supplied by human friends for such birds as may choose to make them their temporary homes.

WHITE-BELLIED SWALLOW.

These Swallows are gregarious, and often assemble in great bands or companies. The wires stretched between telegraph poles are favorite resting places, and often they crowd such perches so that it seems impossible for another one to alight. While insects are their chief food, they are well known to feed late in fall on the ripe fruit of the bayberry.

These birds are a little less than six inches long. Old ones are uniform dark steel blue or green above, and unsullied white below. The tail is very slightly forked. Immature birds are grayish brown above and white below.

The nests are built in the kind of shelters indicated above. They are made of coarse grasses and feathers. From four to seven unmarked white eggs are laid. They are about three quarters of an inch long and more than half an inch broad.

White-bellied Swallows are found throughout North America at large as far north as Labrador and Alaska. They breed from New Jersey and North Kansas northward, and winter from the South Atlantic and Gulf States southward.

The Bahaman Swallow is nearly six inches long, and is of a curious grass green, soft and metallic in character, somewhat iridescent, but not very **Bahaman Swallow.** glossy, on the back. This color shades into bluish green **Callichelidon cyaneoviridis** on the wings, rump, and tail. The entire lower parts and **(Bryant).** sides of the head below the eyes are uniform pure white. The tail is very decidedly forked, the outer feathers being much the longer, and recalling, in a way, the tail of the Barn Swallow.

This bird occurs in the Bahama Islands, where it is resident and a common bird. The writer secured a male bird and saw one other at Garden Key, Dry Tortugas, Florida, on April 7, 1890.

Later, in the summer of 1892 a young bird in its first plumage was taken in the vicinity of Tarpon Springs, Florida.

It seems probable that the birds occur rarely and breed on the Florida Keys and the southern coast of the Peninsula.

The Black-throated Bunting does not occur plentifully in the Coast States of the region we are dealing with, and it is not a noticeable factor in the bird fauna east of the Alleghanies except locally, though **Dickcissel.** it was formerly common in the region from Massachusetts **Spiza americana (Gmel.).** southward.

It is still an abundant bird in the Mississippi Valley, where it breeds from Texas to Ontario. It winters from parts of Mexico and through Central America to Northern South America.

They are birds of the open grass fields, nesting in low bushes or on the ground.

The birds are robust in form and about six inches in length. Adult males have the head and sides of the face slaty gray ; the back is striped with black on a brownish gray ground, and the rump is grayish brown unmarked.

The wings and tail are dusky, and the shoulders reddish brown. There is a yellow stripe over the eye, one on, either side of the throat, and the breast is yellow back to the white belly. The upper throat is white, and below this is a black patch reaching the yellow of the breast.

The female differs in having the head streaked with dusky brown, and lacks the black throat patch, though this feature is indicated obscurely in some individuals. There is also less yellow on the breast. In fall the birds appear more highly colored, there being a suffusion of reddish brown about the head and back.

The nest is made of leaves, plant fibres, and grass, and lined with finer material and some horse hairs. The eggs are pale blue, unmarked, and vary in number from three to five. They are about four fifths of an inch long, and three fifths of an inch in their median diameter.

The Lark Bunting, or White-winged Blackbird, is a finch which is common from Middle Kansas westward, ranging in winter into Texas. It is

Lark Bunting.
Calamospiza melanocorys Stejn.

a bird of the central regions of North America north to our northern boundary, and has been recorded from South Carolina, Long Island, and Massachusetts.

The adult male is black with a large white patch on the shoulders. The female is grayish brown above, streaked with dusky, and having a smaller white wing patch. Below she is white on the belly. streaked on the breast and sides with dusky brown and blackish. The male in winter is not unlike the female in summer.

The Grassquits are small, and generally plain olive green finches above,

Grassquit.
Euethela bicolor (Linn.).

Melodious Grass-quit.
Euetheia canora (Gmel.).

with the under parts varying from black to almost white in color. The two kinds recorded from the Eastern United States have been found on the Florida Keys only once.

They are between four and five inches in length, and are inhabitants of the Bahamas and the West Indies.

This finch is one of our Southern birds, and the male is among the most gaily colored of our native birds. Not quite as large as the Indigo

Bird, to which it is closely related, this bird is between five inches and a quarter and five inches and a half long.

    In the vicinity of New Orleans, where the bird breeds
Painted Bunting. and is common, it is known as the " Pape," from its fine blue
Passerina ciris (Linn.). hood covering the head and sides of the neck. The back is a peculiarly shaded green with a brilliant golden tinge. The rump is deep red, and the wings and tail are dusky with a tinge of deep red. The entire lower parts and eyelids are vermilion. This is the adult male, and the female and immature birds are deep olive green above, the wings and tail being dusky and shaded with the olive green of the back. The lower parts are uniform dull yellowish olive, sometimes almost grayish or whitish, shaded with yellow and dull olive.

    The nesting habits and economy are similar to that of the Indigo Bird, and the bluish white eggs, from three to five in number, are marked with varying shades of brown. They are a little less than four fifths of an inch long and almost three fifths of an inch in their smaller diameter.

    The birds are found in the South Atlantic and Gulf States, ranging as far north as North Carolina and Southern Illinois and Kansas. They winter sparsely in the extreme Southern States and thence south to Panama.

Varied Bunting.   The Varied Bunting, a bird of the Rio Grande Valley
Passerina versicolor (Bonap.). and the Mexican border, has been taken once in Michigan.

    In old fields and pastures where clumps of bushes have sprung up, and in the growth along the fences and walls that form dividing lines you will see Indigo Birds.
Indigo Bunting.   They are small birds, about five inches and a half long.
Passerina cyanea (Linn.). The male is unmistakable in his coat of deep blue, which is most intense on the head and sides of the face and brightest on the back. The wings and tail are dusky or black and the exposed edges of each feather are blue. The female is brownish, unstreaked above and lighter on the under parts, becoming whitish on the belly. The wings and tail are dusky, sometimes showing suggestions of blue on the exposed edges of the feathers. Immature birds are much like the female, and adult males in the fall and winter are heavily washed with reddish brown showing an admixture of blue. The nest is placed near the ground in a fork or crotch generally in dense

thickets. It is made of various plant fibres and lined with similar finer material and horse hairs. The eggs, three or four in number, are bluish white. They are less than three quarters of an inch long and over half an inch in their median diameter.

The Indigo Bird is found in the breeding season throughout the Eastern United States, and north to New Brunswick and Minnesota. It winters in Central America.

The Blue Grosbeak is a bird about seven inches in length. It is a conspicuous bird on account of its rich color, and is generally found in open country where there are growths of small trees and clumps of bushes.

Blue Grosbeak.
Guiraca cærulea (Linn.).

The adult male is deep clear blue, which color is darker about the head and face. The wings and tail are dusky or black, each feather edged narrowly with blue. There are two reddish brown wing bars. The female is grayish brown above, and a paler shade of the same color prevails on the under parts. The wings are dusky with bars much as in the male, but lighter buffy brown. Some individuals show a slight admixture of blue feathers about the head. Immature birds resemble the adult female and adult males in the autumn are much suffused on the back and breast with reddish brown.

The nest is a compact structure rather deep, and composed of plant fibres and grasses. It is placed in low bushes, or in tangles of the stronger tall weeds. The eggs vary from three to four in number and are pale whitish blue. They are about four fifths of an inch long and two thirds of an inch in their smaller diameter.

The birds are found in the Eastern United States from Southern New Jersey southward, but do not seem common, except locally, east of the Alleghanies. West of these mountains they appear regularly from Southern Illinois southward. They are of casual occurrence in New England. They winter in Cuba and Mexico.

The Chewink is rather a large finch, being nearly eight inches and a half in length. The male is glossy black above including the tail and wings, the entire head, throat, and upper breast. The exposed edges of the larger wing feathers are white, as

Towhee.
Pipilo erythrophthalmus (Linn.).

235   TOWHEE OR CHEWINK. ADULT MALE.

are the outer three or four tail feathers on their terminal halves. The black of the breast is sharply defined against the white of the chest, and the belly and feathers below the tail are white. The sides are bright reddish chestnut. The *eyes are red* with black pupils.

The female resembles the male in pattern, but the black regions are replaced by bright snuff brown and the chestnut of the sides is not so bright.

Very young birds have a curious streaked plumage, but may be readily recognized by their general form and the markings of the wings and tail which are much the same as in the adults. The nest is placed on or near the ground. It is built of large dead leaves, grasses, and plant fibres lined with finer material.

The eggs vary from four to five in number, and are white finely and regularly specked with reddish brown. They are almost an inch long and about seven tenths of an inch in their other diameter.

The birds are found in Eastern North America north to Southern Canada, and extend west to the Plains. They breed from Georgia and the lower Mississippi Valley north, and winter from Virginia and Arkansas to Florida and Texas.

The Chewink is a bird of the ground. Where thickets prevail in open places, in brush heaps that are left in clearings in the woods, and in bushy undergrowths of the more open woodlands, you are almost sure to find him. Here his frequently uttered call notes and quick energetic movements at once arrest attention.

This bird is the close ally and prototype of the Chewink in Florida and north along the coast as far as Southern South Carolina. It is a smaller

White-eyed Towhee.

Pipilo erythrophthalmus alleni Coues.

bird, rather less than eight inches long, has less white on the wings, and only the two outer tail feathers with white on their terminal halves. The eyes are *cream color* or *white*, with black pupils. The brown areas in the female are deeper in color than the more northern bird.

This is a bird of the ground, like its congener, but is almost confined to regions of thick scrub palmetto.

In old fields and pastures that are more or less overgrown with bushes, such as are dry, and on the upland where young junipers are beginning to claim the land again for forest, you may find a sparrow a little longer and more delicately built than the Chipping Sparrow. Its general color, how-ever, is more tawny above, and his reddish brown bill is

Field Sparrow.
Spizella pusilla (Wils.).

a marked feature. The bird is about five inches and three quarters in length. It has a reddish brown crown, and the back is of a similar shade, the feathers being striped with finer lines of black and edged with grayish brown. The rump is ash color. There are two whi-tish wing bars on each wing. The sides of the face, back of the neck, a line above the eye, and the upper throat are ash gray. The region about the ears is reddish brown. The lower parts are white washed with buff on the breast, sides, and flanks. The tail and wings are dusky, with whitish gray and faint reddish edgings to the feathers. The sexes are alike, and imma-ture birds are duller in general color, often having a narrow median line of ash gray on the crown.

FIELD SPARROW.

The nest is placed on the ground or in low bushes. It is built of com-paratively coarse plant fibre, some rootlets, and lined with finer material and horse hairs. The eggs are bluish white, with reddish brown specks and markings, principally at the larger end. They vary in number from three to five, and are nearly seven tenths of an inch long and rather more than half an inch in their other diameter.

These birds are found throughout Eastern North America as far north as Southern Canada, and south to the Gulf States. They extend westward to the Plains. They breed from South Carolina and Southern Kansas northward, and winter from Illinois and Virginia southward. There are numerous records of the Field Sparrow in midwinter in Southern New York and New Jersey.

Western Field Sparrow.
Spizella pusilla arenacea Chadb.

This is the prototype of our Field Sparrow, taking the place of that bird on the Plains from Texas to Montana. It is an essentially similar bird, but paler reddish and grays characterize it, and it is a rather larger bird.

It has been recorded from near New Orleans.

Clay-colored Sparrow.
Spizella pallida (Swains.).

This bird is found in the central regions of the United States, ranging from Illinois and Iowa west to the Rocky Mountains. It breeds from Illinois and Nebraska northward, and winters in Southern Texas and Mexico. It has been recorded from North Carolina. It averages rather smaller in size than the Chipping Sparrow, or about five inches and one third in length. It has a general resemblance to an immature Chipping Sparrow. The upper parts are pale brownish ash, the feathers of the back and head being closely streaked with dusky brown or blackish. The crown is divided by an ashy median line, and there are lines above the eyes of the same color. Below the prevailing color is grayish white, washed with faint brownish on the breast and sides, lightest on the belly.

The nest and eggs resemble those of the Chipping Sparrow.

Brewer's Sparrow.
Spizella breweri Cass.

This Sparrow is about five inches and a half long, being a little smaller than the Field Sparrow. It is light grayish brown above, rather indistinctly striped or streaked with dusky. The lower parts are grayish white lightest on the belly.

It is a western bird found from the Rocky Mountain region westward. Northward it ranges to Montana and British Columbia, and southward it is found to Cape St. Lucas and in Mexico. It breeds throughout its range in the United States, and has been once recorded from the State of Massachusetts.

When the great troop of birds have passed by and the first snows come, there still remains a brownish sparrow to glean a living during the severest storm and cold.   Like the Snowbird, the Tree Sparrow is gregarious, and is, in fact, often associated with that bird, particularly in bushy open fields, and sometimes about the house.

Tree Sparrow.
Spizella monticola (Gmel.).

This Tree Sparrow is larger than his congeners, the Chipping and Field Sparrows, being about six inches and a quarter long.   The prevailing colors above are different shades of brown, and below grayish white.   There is a clearly defined reddish brown crown, and a streak of the same color behind the eye.   A line above the eye is bluish gray, which color prevails on the sides of the face.   The upper part of the bill is *black*, the lower part yellow at the base and dusky at the tip.   The back of the neck is grayish brown. The back is variegated with broad streaks of black, and narrower ones of buff and reddish brown.   The rump is light brown with a grayish tinge.   There are two white bars on each wing.   The wings and tail are dusky, the exposed edges of the feathers being edged with whitish gray, light grayish brown, and reddish brown.   The throat is grayish, changing to lighter gray on the breast, which is marked with *a single obscurely defined dark spot*.   The belly is white, and the sides and flanks are washed with light grayish brown.

The nest is placed on the ground or near it.   It resembles that of the Field Sparrow in its material and construction.   The eggs vary in number from three to five.   They are much like those of the Chipping Sparrow in color and markings, but larger, being about three quarters of an inch long and nearly three fifths of an inch in their other diameter.

The birds breed in Labrador and the Hudson's Bay region.   They range south in winter to the Carolinas, Kentucky, and Eastern Kansas.

This is another rather large finch, being about six inches and a half long, that is found commonly in the interior of North America, and which appears as a straggler on our Eastern Coast States.   There are various records of its occurrence from Massa-chusetts, Long Island, New York, New Jersey, the vicinity of the city of Washington, and Florida.   In the latter locality I have twice met with it in the fall and winter.   Its regular habitat is from Texas to Southern Ontario, from the States of the Mississippi Valley north of Alabama on the east, and west to the Plains.

Lark Sparrow.
Chondestes grammacus (Say).

Its characteristics are a deep chestnut crown divided by a white median stripe, and bordered by a similar streak over the eye. There is a black line in front of the eye reaching to the base of the bill, and extending a short distance back of the eye. The region about the ears is deep chestnut, and that immediately below the eye whitish. Below this is a defined white stripe reaching from the bill to back of the ear patch, and this is divided from the throat by a clear black line becoming an undefined patch on the sides of the neck. The rest of the upper plumage is grayish brown, streaked with dusky on the back. The tail and wings are dusky, the outer tail feathers are tipped with white. The entire under parts are white with a small distinct black spot on the breast.

The birds nest generally on the ground, but frequently in low bushes. The nest is built of coarse grass, lined with finer grass, fine root stems, and horse hairs. The eggs vary in number from three to five and are white, marked in an irregular way with spots and blotches of shades of dark brown or black. They are about four fifths of an inch long and three fifths of an inch in their other diameter.

This group of Sparrows are small in size, and of rather robust build. They frequent grass fields and prairies, and assimilate in color so closely with their environment, that this factor, together with their skulking, secretive habits, leads to their often being overlooked where they are not uncommon. One rarely sees any of these birds alighted, and like some of the game birds and rails, they lie so close in the long grass that it is often very difficult to flush them. Even when flushed the momentary glance one gets is unsatisfactory and fails to identify the bird save to the experienced observer.

The more common of the trio in the fields of the Atlantic seaboard is the Yellow-winged or Grasshopper Sparrow, and this is also a somewhat

Grasshopper Sparrow.
Ammodramus savannarum
passerinus (Wils.).

larger bird than the other two. The prevailing color of this bird is light buff. The crown is dusky, divided by a central line of ashy buff and bounded by a stripe above the eye of similar color. The region in front of *the eye is brownish orange.* The feathers of the rest of the upper parts are dusky reddish brown or black, bordered and margined clearly with ashy or buff, the reddish brown ones prevailing on the nape and rump, the dusky and black ones on the back. The wings and tail are dusky, the latter composed of *narrow pointed feathers* of about equal length. The *bend* of the *wing* is *bright yellow*, the feathers of the shoulders olive with a buff or yellow tinge, and the smaller wing feathers are variegated with white, buffy and reddish brown. Below the birds are brownish buff, with little or no streaking, becoming white on the belly. The birds are about five inches long.

The Yellow-winged Sparrow is found in North America, east of the Plains from the Gulf States to Massachusetts and Minnesota, breeding throughout this region in suitable localities. In winter it ranges from Florida to Central America.

Henslow's Bunting is an appreciably smaller bird than the Yellow-winged Sparrow, being less than five inches long. It seems to prefer damper meadow grounds, but is also found in dry grass fields sometimes

**Henslow's
Sparrow.**
Ammodramus henslowii
(Aud.).

in close association with its congener. It is a *darker*, *greener* bird, has the *yellow mark in front* of the eye and on *the edge of the wing-bend*. The tail is composed of narrow sharp-pointed feathers, the outer ones being much the *shorter and grading up to the middle ones*. The *green* is *most obvious* on the *back of the neck*, where *each feather is streaked with a narrow black or dusky line*.

The feathers of the back have their centres black or dusky, bordered by chestnut, and edged with ashy. There is a black or dusky line behind the eye, one from the corner of the mouth, and another defining the throat. The under parts are whitish, washed with buff on the *sides, flanks*, and *breast, where they are clearly streaked* with narrow stripes of black or dusky. Henslow's Sparrow is found in the breeding season locally from Virginia and Missouri north to New Hampshire and Southern Ontario. It winters from south of its breeding range to the South Atlantic and Gulf States.

Leconte's Sparrow, the smallest of the trio, is a bird of more western distribution, breeding from Dakota and Minnesota north into Manitoba. It

Leconte's Sparrow.
Ammodramus lecontei
(Aud.).

migrates through Iowa, Illinois, and Kansas to South Carolina, and the Gulf States from Western Florida to Texas. About the size of Henslow's Sparrow, it is a bird of a general tawny reddish color above. The crown is dusky, divided by a central buffy line, and bordered by a broad buffy stripe above the eye. The back of the neck is reddish brown, each feather has a small black central spot giving the effect of dusky streaks. The feathers of the back have black or dusky centres, surrounded by areas of reddish brown, changing to buff and whitish at their edges. The tail is brown with a reddish tinge. It is composed of narrow pointed feathers, the outer ones being much the shorter. The breast, sides, and flanks are tinged strongly with buff more or less streaked with dusky. The belly is white. It has no yellow about the eye or on the bend of the wing, thus readily distinguishing it from the other two birds of this group. Leconte's Sparrow is generally less than five inches long.

The nesting of these three birds is essentially alike. They build nests of grasses and plant fibres on the ground in grassy places. The eggs are white in ground color speckled more or less with varying shades of brown, from pale reddish to almost black. They vary but little in size, being about three quarters of an inch long by half an inch in their smaller diameter, those of the Yellow-winged Sparrow being rather rounder as compared with the other two kinds.

The Savanna Sparrow is a bird of robust build, about five inches and a half long. In color it is grayish brown above with the broad dusky or blackish streaks of the centre of the feathers of the back contrasting strongly against their grayish brown edgings. There is a light yellow spot in front of the eye, frequently extending above it, and the same color is noticeable on the bend of the wing. The tail and wings are dusky. The exposed edges of the wing feathers are edged with grayish buff. The tail has the outer web of the feathers white or grayish white. The under parts are white or grayish white, much streaked with dusky and dark reddish brown, the feathers on the breast having arrow-shaped tips of these colors.

Savanna Sparrow.
Ammodramus sandwichensis savanna (Wils.).

The nest is built, on the ground, of grasses and mosses, and lined with finer plant fibres, grasses, and hair. The eggs are faint bluish white, much speckled and frequently washed with bright reddish brown. They are nearly four fifths of an inch long and less than three fifths of an inch in their other diameter. These birds are found in parts of Eastern North America throughout the year. They breed sparingly from Northern New Jersey, and plentifully from the Northern United States to Labrador and Hudson's Bay. They winter from Southern Virginia and Illinois, southward to Cuba and Mexico.

These are true field birds and may be observed in numbers where they breed and during their migrations in open grassy places or along country

Across the Fields.

highways.   They sometimes lie close, particularly during the breeding season,
when they rise almost at your feet, whirring away like some miniature game
bird.   Alighted again, they sneak off with such rapidity through the grass

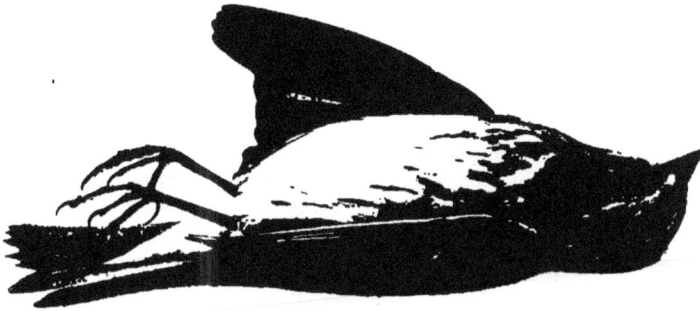

SAVANNA SPARROW.

which conceals them that though surely expecting to easily see the bird near
at hand, you fail till he is unexpectedly flushed again at considerable distance from where you marked him down.

The Ipswich Sparrow is in effect a pale grayish brown bird, about six
inches and a half in length.   It is a maritime bird, generally found close to
the sea.   The sand dunes of the coast with their scanty
tufts of beach grass are the regions in which to find these
birds, during their migrations and in winter.   They
strongly resemble the Savanna Sparrow and may be readily mistaken for that
bird.

Ipswich Sparrow.
Ammodramus princeps
(Mayn.).

The birds are known to breed on Sable Island, Nova Scotia.   They
migrate along the Atlantic coast and winter regularly as far south as Virginia
and rarely to Georgia and points farther south.

The Ipswich Sparrow is pale grayish above; the top of the head and the
back are streaked with pale and dusky brown.   The back of the neck and

rump have little or any of this marking. There is a median gray streak, dividing the crown and a lighter stripe above each eye defining it. Usually there is a dull yellow or pale sulphur area in front of the eye, and the bend of the wing almost always shows the same shade. Below the birds are white washed with pale brown on the sides and flanks and sometimes on the breast. The breast, sides, and flanks are streaked with dusky brown.

The nesting is similar to that of the Savanna Sparrow and the eggs are much like those of that bird. They are more than four fifths of an inch long and three fifths of an inch broad.

Lapland Longspur.
Calcarius lapponicus (Linn.).

The Lapland Longspur is a boreal bird, breeding in the northern portions of the Northern Hemisphere. In the winter migrations they regularly reach the northern borders of the United States. On the prairies of the interior they are especially abundant during the winter. Sometimes they appear in Northeastern America, as far south as the Middle States, and they have been recorded as far south as South Carolina.

The birds are about six inches and a quarter long, and the males are somewhat longer than the females. In the breeding season old males are black on the head, neck, and breast. This area is broken by a broad buff or white stripe behind the eye, prolonged downward and backward along the sides of the breast. The back of the neck is reddish brown. The back is black streaked with light buffy brown. The tail is dusky with some white on the webs of the outer feathers. The belly is white and the sides and flanks are streaked with black. In winter the black on the head appears only on the crown, in the region about the ears, and somewhat on the lower throat and breast. The rest of the head, the back, and sides of the neck, the upper throat, and much of the breast are dull buffy brown light in tone. The back is streaked with black, reddish brown, and varying shades of buff. The under parts are white and the sides and flanks streaked black and brownish buff.

The female in the breeding season is similar in color to the male in winter but with the black regions more restricted and broken and the back of the neck streaked with black. In winter the females have little or no reddish brown on the back of the neck and the colors throughout are obscured by brownish.

The birds build nests of moss and grasses on the ground. From four to six bluish white eggs are laid much washed and obscured by grayish brown. They are little more than four fifths of an inch long and about three fifths of an inch broad.

This Bunting or Longspur is during the cooler portions of the year a characteristic bird of the prairies, from Illinois to Texas, and extends westward over the Great Plains of the interior of the United **Smith's Longspur.** States. In summer they range far north, breeding to the *Calcarius pictus (Swains.).* Arctic Coast.

The birds are nearly six inches and three quarters long. In the summer plumage the adult male has a black head. There is a white line over the eye, a spot on the back of the neck, and one on the region about the ears is of the same color, and there is a large white patch on the wings. The back is black streaked with buff. The tail is dusky, with the two outer feathers nearly white. A band on the back of the neck and the entire lower parts are yellowish buff. In winter, the defined markings of the male are replaced by dusky streaked with brown, and the throat and chest are streaked with dusky and are lighter buff than in summer. Otherwise the birds are much as at that season. The adult female resembles the winter male.

The nest is, like that of its congeners, built on the ground, of mosses and grasses, and lined with finer material of the same sort. The eggs are bluish white speckled with reddish brown, varying to nearly uniform dark brown. They are over four fifths of an inch long, and about three fifths of an inch in their other diameter.

There are two other birds closely allied to Smith's Longspur and the Lapland Longspur that have a place in the bird fauna of Eastern North America. One is the Chestnut-collared Longspur, which **Chestnut-collared** is a bird of the Great Plains and the interior of North **Longspur.** America. It breeds from the western part of Minnesota *Calcarius ornatus (Towns.).* north to the Plains of the Saskatchewan region.

In the adult male in summer the crown, entire breast, and upper belly are black, as is a line back of the eye and a crescent shaped mark on side of the head. There is a bright chestnut band on back of the neck. The rest of the upper parts are brownish gray with darker streaks. The throat and sides of the head, the lower belly and feathers below the tail are white.

This bird has been recorded from Long Island and Massachusetts as a straggler.

McCown's Longspur breeds from northwestern Kansas north through
McCown's Long- Nebraska to the Plains of the Saskatchewan. In winter
spur. it is found as far south as Texas and Northern Mexico,
Rhynchophanes mccownii
(Lawr.). and is of casual occurrence in Illinois.

The male in summer has the crown black and there is a black crescent shaped mark on the breast. The shoulders are reddish brown. The remainder of the upper parts are gray or brownish, streaked with dusky; the tail is white, except the two middle feathers, and each white feather is tipped broadly with black. The lower parts, except the chest, are grayish white.

The predominating impression that one gets of a Snow Bunting is that of a *white* bird, but closer observation reveals very distinct and marked areas
Snow Bunting. of browns and blacks, varying with the seasons.
Plectrophenax nivalis The Snow Bunting is a strongly built, closely feathered
(Linn.). bird, calculated to withstand severe cold and storm. The
birds are rather more than six inches and three quarters long.

In summer the males are white, except the back, the shoulders, and the

SNOW BUNTING.

inner tail feathers, and the terminal half of all of the larger and part of the smaller wing feathers, which are black. The female, at this season, has the entire upper parts streaked with black. The black parts of the wings in the male are replaced by dusky.

In the winter there is a general suffusion of rusty brown over the upper parts, the black of the back showing through and giving a streaked appearance to that part. The wings and tail are similarly suffused. The under parts are white, except on the breast, sides, and flanks, which are washed with rusty brown.

The birds nest on the ground, building a structure of mosses, grasses, and various plant fibres and some feathers. The eggs are bluish white, varying from sparsely speckled to heavily marked and washed with varying shades of brown. They are about nine tenths of an inch long and rather more than three fifths of an inch broad.

These birds are found throughout the northern parts of the Northern Hemisphere, breeding in the Arctic regions. They migrate into the northern United States in the fall and winter and are found generally in large flocks as far south as New Jersey and Virginia on the coast and Southern Illinois and Kansas in the interior.

There are two kinds of Redpolls that visit the United States in winter, and which breed in the northern parts of the continent. One kind is represented by two, and the other by three geographical races, which will be noticed in detail later.

In general appearance all these birds are similar, having as characteristic features in their adult plumage deep pinkish red crowns, a more or less defined dusky throat patch, a suffusion of rosy color on the whitish rump and breast, extending in individuals to the sides. The wings and tail are dusky. The upper parts are streaked with dusky and yellowish browns, except in the region indicated, and the belly is white. The sides and flanks are streaked with dusky. Female birds have little or no rosy shading on rump and breast.

They vary somewhat in size, but are all close to five inches in length.

They breed in low bushes, or on the ground in grass tussocks. The nest is made of grasses, dry mosses, and like material, and is lined with plant down, feathers, and hair. The eggs are bluish or greenish white, spotted with reddish brown. They vary in number, from four to six, and are more than three fifths of an inch long, and about half an inch broad.

The most common of the group in the United States is the Redpoll.   It is about five inches long, the back is, on the whole, dark, the feathers have tawny yellowish brown margins.

Redpoll.
Acanthis linaria (Linn.).

The Redpoll breeds far north in the Northern Hemisphere, and migrates irregularly south in North America, as far as Virginia and Illinois.

Holböll's Redpoll is the race of this bird that breeds in the northern portions of Europe and Asia, and also in parts of Alaska.   It has been recorded once in Massachusetts and once in Quebec.

Holböll's Redpoll.
Acanthis linaria holbœllii
(Brehm).

It is a larger bird than its ally, the Redpoll, with a proportionately longer bill.

The Redpoll that breeds in Southern Greenland is also a race of the common Redpoll, and is known as the Greater Redpoll.   It is the largest of the *linaria* group, being about five and a half inches long,

Greater Redpoll.
Acanthis linaria rostrata
(Coues).

and is rather darker in color than its congeners.   It migrates irregularly to the Northeastern United States in winter, to New England, and Northern Illinois.

The Greenland Redpoll, which breeds in Northern Greenland, and ranges south through Labrador in winter, is the second of our two kinds of

Greenland Redpoll.
Acanthis hornemannii
(Holb.).

Redpolls.   It is the largest of the Redpolls, being about six inches long.   The back is dusky grayish brown, light as compared with *A. linaria*, and the feathers with whitish or grayish margins.   There is little or no streaking on the rump and less streaking on the sides and flanks.

The Hoary Redpoll is the race of the Greenland Redpoll that sometimes crosses our borders in winter.   It breeds in the Arctic regions of the

Hoary Redpoll.
Acanthis hornemannii ex-
ilipes (Coues).

Northern Hemisphere.   It is smaller than its congener, the Greenland Redpoll, being about five inches and a quarter long and of even lighter general color, whence its name.

The Starling of Europe has been introduced in the vicinity of New York, and with apparent success, as the birds have increased in numbers and

Starling.
Sturnus vulgaris
(Linn.).

may be seen in small flocks in the different parks and at points in the recently annexed portions of Westchester County. It has also been recorded as an accidental straggler in Greenland.

The birds are about eight inches and a half long, and are of a general dark metallic green or purple color. Above each feather is tipped with a buff spot, and below only those of the sides are similarly marked. The nest is made of grasses and sticks and lined with finer material. It is placed in some convenient crevice in the eaves of buildings, or at times in hollows in trees. The eggs vary from four to six in number, and are pale blue in color. They are rather less than an inch and a fifth long and nearly nine tenths of an inch broad.

The Crow is a bird about nineteen inches and a half in length and clear black, with blue and purplish sheen throughout. The upper parts, wings

American Crow.
Corvus Americanus (Aud.).

and tail, are more brilliant than the under parts. The sexes are alike, and young birds are dull black or dusky, having none of the sheen till after the first moult. The nest is a bulky affair, made of sticks and lined with strips of bark, dried cow and horse dung, grasses, and moss. It is placed in the crotch or fork of a tree usually more than thirty feet from the ground. Four to six eggs are laid. These are greenish or bluish green, heavily marked with varying shades of dark brown. This is the usual color, but the eggs vary much and are sometimes white or pale blue, with very few markings. They are an inch and seven tenths long and less than an inch and a fifth broad. The Crow is found throughout North America from the Fur Countries to Mexico. It is of local distribution in the West and probably does not occur in Florida, being there represented by the Florida Crow. It winters from the Northern United States southward.

One of the distinctive birds of the open, found in old fields and pastures and meadows, who loves the prairies of the South and West equally

Meadowlark.
Sturnella magna (Linn.).

well, is the Field- or Meadowlark. A bird that ranges over a vast area, which does not migrate very far from its

250 AMERICAN OR COMMON CROW.

breeding place, and at many points not at all, it is marked by a certain sort of provincialism that is not to be passed by unnoticed.

Before the days when facilities for travel began to increase, before the posthorse was the synonym for rapid transit, travel was full of difficulties and trials not to be undertaken without due deliberation. Only stern necessity of some sort forced anyone on prolonged journeys and men were on the whole practically stationary. Their intercourse was limited to the vicinity of the place in which they were born. One of the results was a diversity of

MEADOWLARK.

dialects of those who spoke the same general language, so that men speaking German, or French, or even English, from different regions of their respective lands, found it not only difficult but often impossible to understand the language of their fellow countrymen. Birds' songs, supplemented by their various call notes, are undoubtedly their language. They learn it from one another. Orioles, Robins, Wood Thrushes, Jays, and other birds which I have reared from young birds, birds that were not educated by their parents, all sing,—but no song that their friends in the forest would recognize or understand—no song by which a trained ornithologist would recognize them, without seeing the singer. And should you hear the song of the Meadowlark, say at Denver, near New York, or at any point in Florida, I feel sure you would never recognize it as the song of the same bird. They have not travelled far, provincialisms in song and call note have sprung up and have been adopted and perpetuated till now, at the several extremes of its habitat, this practically non-migratory bird has developed his own dialects, which grade into one another, but which in their extremes have no resemblance.

The Meadowlark is a bird nearly eleven inches long. Its prevailing tone above consists of blacks, browns, and dry grass colors, difficult to describe in detail, but forming on the whole a general streaked effect. There is a narrow line of buff dividing the brown crown into two distinct halves, each of which is bordered by a broader buff stripe, which becomes bright yellow just above the eye and reaches in front of it to the bill. The eyelids, the sides of the face and the region about the ears are grayish. There is a distinct dark brown stripe beginning back of each eye and extending back on the sides of the head. The throat and breast, chest and belly are bright yellow. This is broken on the upper breast by a broad black crescent shaped mark. The middle tail feathers and the exposed surfaces of the wings are colored much like the back but are *brokenly barred* transversely instead of longitudinally. The outer tail feathers are partly white on both the inner and outer webs. The sides and flanks are buff deeply and clearly streaked with dusky brown markings. In the winter the birds are duller and browner having a general suffusion of that color effected by the brownish edging to the feathers. The yellow breast and the black crescent are obscured in the same way.

The nest is built on the ground concealed by the tall grass of early summer. It is made of grasses throughout and is sometimes partly roofed with a semi-down of the same material. From four to six white eggs speckled with reddish brown are laid. These are about an inch and an eighth long and four fifths of an inch broad. The birds are generally distributed throughout Eastern North America from Southern Canada southward and west to the Great Plains. In the extreme north they are migratory but begin to winter from Massachusetts and Illinois southward and it seems probable that birds from the latitude of New Jersey south are resident.

Western Meadowlark.
Sturnella magna neglecta (Aud.).

This is the geographical race of the Field Lark that is found throughout Western North America from British Columbia southward. They are found eastward regularly to Kansas and Texas and less commonly in Wisconsin and Illinois. They are rather larger birds than their allies. The prevailing tones of color above are lighter, and the barring on the wings and tail is distinct and not broken or interrupted as in the representatives from the East. The yellow of the throat reaches farther back on the sides of the neck almost to the region about the ears.

The male Cowbird is about eight inches long and has a hood of fine dark snuff brown extending over the head, neck, and breast. The rest of the plumage is lustrous metallic black with an iridescent sheen varying from blue to greenish. The female is a plain grayish brown bird with this tone lightest below often becoming grayish white on the throat.

Cowbird.
Molothrus ater (Bodd.).

From its method of nesting, or rather lack of it, for so far as known these birds do not build but are parasitic, laying their eggs in other birds' nests, the Cowbird has called down maledictions on its head. This may be to an extent warranted, but the fact that the great laws of nature have developed a necessity for such a bird seems to bespeak for it at least patient and careful consideration. There are few if any unmixed evils allowed to survive in the great struggle for existence, but the good results are not always patent to even the most careful student.

Many of our birds prey directly on each other but we have discriminated and balanced the accounts of a few that *seem the worst*, notably Hawks, Owls, Jays, and Crows, and on the whole have concluded that there is a large balance of good to their credit after charging up all apparent evil. It is more than probable that even in the case of the despised and maligned Cowbird this is true. Perhaps the cows and beasts of the field, with whom these birds associate on apparently good terms, would prove good advocates for their small friends, telling of myriads of insects that are kept in check by their efforts. After watching them for years I cannot but believe this.

The Cowbird's egg is white profusely and closely spotted with brown shades that give it a gray effect. They are nearly nine tenths of an inch long and over three fifths of an inch broad. You will find such eggs in many kinds of small birds' nests, notably some of the Warblers, Sparrows, and Vireos.

The birds range from Texas into British America as far north as New Brunswick and Manitoba, breeding throughout this area and wintering regularly from Southern New Jersey and Illinois southward.

The Bobolink is almost a synonym for summer. He comes to us on the flood tide of life in gay coat and with a marvellous song which are almost as evanescent as the first color and texture of the leaves. He is here and gone with the long June days, and to such as have known him is associated with them as one of their greatest charms. Early July finds him silent, and moulting he becomes an inconspicuous bird sought by the gourmand as a table delicacy and known by the name of " Reedbird," " Ricebird," or " Pink." He is a bird about seven inches and a quarter long.

Bobolink.
Dolichonyx oryzivorus
(Linn.).

The male Bobolink is almost as gay in his parti-colored dress as the flower decked meadow over which he hovering sings,—a song that only a poet can interpret.

> " What the fun was I could n't discover ;
>   Language of birds is a riddle on earth ;
> What could they find in white-weed and clover
>   To split their sides with such musical mirth ?
>
> " Was it some prank of the prodigal summer,
>   Face in the cloud or voice in the breeze,
> Querulous cat-bird, woodpecker drummer,
>   Cawing of crows high over the trees ?
>
> " Still they flew tipsily, shaking all over,
>   Bubbling with jollity, brimful of glee,
> While I sat listening, deep in the clover,
>   Wondering what their jargon could be.
>
> " 'T was but the voice of a morning, the brightest
>   That ever dawned over yon shadowy hills ;
> 'T was but the song of all joy that is lightest—
>   Sunshine breaking in laughter and trills.
>
> " Vain to conjecture the words they are singing ;
>   Only by tones can we follow the tune
> In the full heart of the summer fields ringing,
>   Ringing the rhythmical gladness of June."
>
> —*Christopher P. Cranch.*

He has the entire lower parts clear black, which extends to the sides of the face and neck and to the top of the fore part of the head. There is a fine buff cap on the back of the head defined by the black of the back of the neck and upper back. This black area is more or less striped with buff. Back of it the back is grayish, shading into clear white on the rump and feathers above the tail. The tail is black and its feathers are sharply pointed. The wings

BOBOLINKS. MALE AND FEMALE IN BREEDING SEASON.

BOBOLINK'S NEST AND EGGS.

are black, with the exposed edges of the feathers margined with buffy white. There is a large patch of white on the shoulders.

The female is decidedly sparrow-like in appearance, dusky brown above, streaked with buff. There is a median line of buff on the head dividing the dusky crown, and an obscure dusky line beginning back of the eye, defining a broad buff stripe above the eye. The lower parts are pale buff, lightest on the belly. The sides and flanks are washed with olive gray and indistinctly striped with dusky. The wings and tail are dusky brown having the exposed edges of the feathers buffy.

All the Bobolinks in the fall are very similar in appearance to female just described, but the adult males have a great admixture of yellowish olive and the buff is clearer.

The birds breed in nests of grass built on the ground in the grass.

The picture on the opposite page is reproduced from a photograph of a nest taken in its original undisturbed position in the grass. The eggs are bluish white, speckled and marked in zigzag lines with dark brown. They are decidedly oriole-like in appearance. They are more than four fifths of an inch long, and a little over three fifths of an inch broad.

The birds breed from Southern New Jersey north to Nova Scotia in suitable localities. They range west to Montana and Utah, and winter in South America, migrating through Florida and the West Indies to that point.

There are two kinds of Horned Larks represented in the bird fauna of this region : the Horned Lark and the Prairie Horned Lark.

The Horned Lark is a bird about seven inches and three quarters long. It is found in the United States as a fall and winter visitor, rather local in its distribution and at this season gregarious. The pre-

Horned Lark.
Otocoris alpestris (Linn.).

vailing color of the male bird on the upper parts is a peculiar shade of pinkish brown, light in tone, more or less washed with cold gray. There is a band of yellowish white across the forehead which extends backward as a stripe above the eye and on the sides of the head. This marking is defined by a black area of similar shape just inside of it, the portion above and on the sides of the head forming the so-called horns. The region in front of the eye is black, and broadens below the eye so as to be triangular in shape. The throat is sulphur yellow, and there is a black crescent shaped mark on the breast. The remainder of the lower parts are white, the sides and flanks are washed with pinkish brown, like the upper parts, but rather lighter. The wings are grayish brown, much like the back, and more pinkish, brown on the shoulders. The middle tail feathers are dusky, broadly edged with a color similar to that of the back. The other tail feathers are black or dusky, and the outer ones have considerable areas of white.

HORNED LARK.

The female is not as bright as the male, nor are the markings as clearly defined. In winter the black markings are somewhat obscured by the white edgings of the black feathers.

The Horned Lark is found in Northeastern America, breeding in Greenland, Newfoundland, and the Hudson's Bay region. In winter they go as far south as Carolina and Illinois.

Prairie Horned Lark.
Otocoris alpestris praticola Hensh.

The Prairie Horned Lark is half an inch shorter than its congener and proportionately smaller. It differs in having the regions about the ears white, and the throat pale yellow, frequently almost white. In general tone it is a paler bird.

It is a bird that has changed its habitat with the settlement and clearing of this country. It formerly was confined to the prairies of the interior, but now breeds and is resident from the Upper Mississippi Valley eastward through New York to Western Massachusetts and Long Island.

The nesting of the Horned Lark and the Prairie Horned Lark is essentially the same. They breed on the ground much after the manner of some of the sparrows, the nest being sunk in the surrounding ground or turf. It is built of grasses and weed stalks. The eggs are pale buff or olive thickly speckled with brown and cinnamon of varying shades. Those of the Prairie Horned Lark are rather more than nine tenths of an inch long and about two thirds of an inch broad. The eggs of the Horned Lark are slightly larger.

Skylark.
Alauda arvensis Linn.

The Skylark of Europe and Asia has been recorded as an accidental straggler in Greenland and Bermuda. Efforts have also been made to introduce it, at sundry times and in various places, in the United States. So far as known these attempts have been, on the whole, unsuccessful.

The Skylark is about seven inches and a half long. It is light brownish above, streaked with black, noticeably on the back. There is a whitish stripe above the eye, and the general color of the lower parts is white; this color shades into reddish buff on the breast, which is distinctly striped with black. The sides and flanks are more or less buffy streaked with dusky.

The birds nest in grassy fields or meadows, on the ground. The eggs are dull white, buffy or brown in tone, thickly speckled with brown. They are nearly nine tenths of an inch long and two thirds of an inch broad.

Alder Flycatcher.
Empidonax traillii alnorum Brewst.

The Alder Flycatcher is an ally of the Traill's Flycatcher of the Western States. It is found during its migrations throughout Eastern North America as far west as Michigan, and north to New Brunswick. It breeds from Northern New England northward, and winters in Central America.

It is the *largest* and *brownest* of the birds of this group, being at least six inches long.

The upper parts are *olive brown* with a slight greenish tone. The under parts are grayish white, darker on the breast and sides, and faintly yellowish on the belly. The throat is pure white. The wing bars are grayish brown.

This bird frequents, during the breeding season, open bushy fields, where alders grow along streams. The nest is built in some bush or sapling near to the ground. It is generally placed in a fork or crotch, and is built of rather coarse plant fibres, lined with finer material of a like nature. The eggs in color and markings closely resemble those of the Acadian Flycatcher, but are a little larger.

Traill's Flycatcher is the western form of the bird just considered, occurring throughout Western North America north to the **Traill's Flycatcher.** Fur Countries, and ranging east into Ohio, Illinois, and Empidonax traillii (Aud.). Michigan. It is very similar to the Alder Flycatcher, but is somewhat browner, and has a comparatively larger bill.

The Kingbird is a bird usually seen along the road, perched often on the fence or on some favorable dead limb. He likes open places where a good view of his passing insect prey may be obtained. His in-
Kingbird. stincts are sportsmanlike and he is notably brave in de-
Tyrannus tyrannus (Linn.). fending himself, his family, and his chosen domain from the intrusions of larger birds such as Hawks and Crows. Not content with

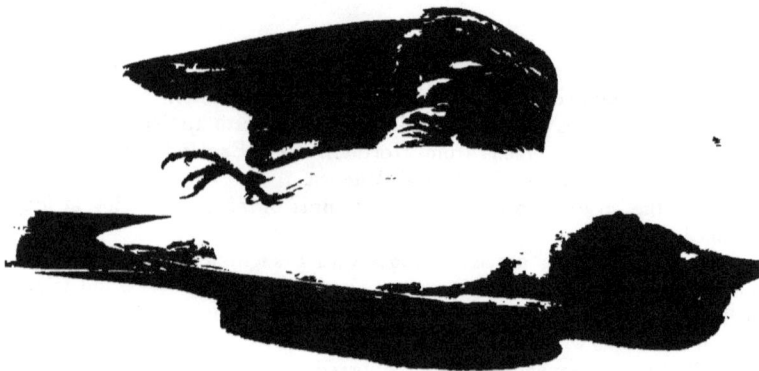

KINGBIRD.

simply putting the enemy to rout he pursues him well away from the vicinity of his nest or perch, driving him often out of sight and then returning satisfied.

Kingbirds are about eight inches and a half long. They have large heads with powerful pointed and hooked bills very broad at the base. Their general color above is grayish slate, which shades into dark lead color on the top of the head and on the feathers above the tail.

There is a bright reddish orange crown patch which is covered and hidden by the dark feathers of the top of the head, save when the whole crest is erected. The under parts are white washed faintly on the breast and more heavily on the sides and flanks with lead color. The tail feathers are black with whitish tips.

Young birds of the year lack the crown patch, but are otherwise much like the adults save for a general suffusion of pale buffy.

The nest is a well made structure of various weed and grass stems and moss and lined with finer grasses, plant down, and the like. It is saddled on a fork near the end of a limb from ten to twenty-five feet from the ground.

From three to five eggs are laid. These are white spotted and specked with dark brown. They are rather less than an inch long and nearly three quarters of an inch broad.

The Kingbird is found throughout North America, north to New Brunswick and Manitoba. It breeds through most of its North American range and winters in Central America and South America. It is much more common east than west of the Rocky Mountains.

The Gray Kingbird has the upper parts clear ashy gray throughout; the feathers of the head conceal an orange colored patch in the middle of the

Gray Kingbird.
Tyrannus dominicensis
(Gmel.).

crown ; the wings and tail are dusky ; the under parts are white, tinged on the breast and sides with grayish. The bird is rather larger than the Kingbird, about nine inches in length. In general habits the two birds are similar, though the Gray Kingbird seems more maritime in its tastes, being seldom found commonly far from water and apparently preferring the vicinity of the sea.

The nest resembles that of the Kingbird, and is placed in like situations. The eggs are buffy white, and spotted much like those of the Kingbird with dark umber brown. They are rather more than an inch long and less than three quarters of an inch broad.

The Arkansas Flycatcher is a western bird that is distributed over the area from the Great Plains to the ￮Pacific Coast. It ranges north to British Columbia and south through Western Mexico to Central America.

Arkansas Kingbird.
Tyrannus verticalis Say.

This bird has been recorded as an accidental straggler in Iowa, Maine, New York, New Jersey, and in the District of Columbia. The Arkansas Flycatcher is about the same size as our Kingbird and is similar in form. The head, neck, and breast are dark lead color shading into whitish on the extreme upper throat and the region about the cheeks. The wings are light brown of a grayish tone. The tail is black, each feather is tipped with pale brownish gray, and the belly and sides are bright yellow.

In general habits these birds closely resemble our Kingbird and the nesting and eggs are very much like those of that bird.

The Scissor-tailed Flycatcher is another accidental bird in Eastern North America, where it has been recorded at Key West Florida, at Norfolk Virginia, in New Jersey, Connecticut, Ontario, and near Hudson's Bay.

Scissor-tailed Flycatcher.
Milvulus forficatus (Gmel.).

Its habitat is from Southern Kansas and Southwestern Missouri through the Indian Territory, Oklahoma, and Texas, south through Eastern Mexico to Costa Rica. It has the top of the head ashy gray shading into light bluish gray on the upper parts, the back being tinged with vermilion. The *tail feathers* are mainly *white*. The lower parts are white, washed with gray on the breast and on the sides and flanks with reddish salmon color. The feathers under the wing and a patch concealed by the gray of the crown are bright scarlet. Adult birds are about fourteen inches long, and the forked tail is about nine inches in length.

The birds build nests in trees not unlike that of the Kingbird and of similar material. From three to five white eggs are laid. They are sparsely but definitely spotted with brown and lilac. They are about nine tenths of an inch long, and they are about two thirds of an inch broad.

The Fork-tailed Flycatcher is a bird of tropical America, which occurs regularly north as far as Southern Mexico. It has been recorded as an accidental bird in Mississippi, Kentucky, and New Jersey.

Fork-tailed Flycatcher.
Milvulus tyrannus (Linn.).

These birds are about thirteen inches and a half long.

Adults have the back light bluish gray. The top of the head and the long tail (about nine inches) are black. There is a patch in the black of the crown much as in the Kingbird, which is pale lemon yellow. The outer tail feather is edged with white. The *lower parts are pure white.*

Its salient characteristic is its *long forked tail, black in color,* the outer feathers being edged with white.

Nighthawk.
Chordeiles virginianus
(Gmel.).

The Nighthawk is a bird in many respects resembling a Whip-poor-will. In form he is much like that bird, but his colors and markings are very different. It is when his life economy is looked into that we find a wide divergence, both as to habits in general and particularly as to the localities the birds affect.

The Nighthawk seeks no dusky wood shade, but is a bird of the open, preferring even in wooded country, clearings, and the more scattered scrubby parts. The bird is about ten inches long.

NIGHTHAWK. ADULT MALE.

The upper parts of the male Nighthawk are black, thickly marked in an irregular way with white and buffy, giving a general grayish tone to the bird. The feathers of the wings and tail are dusky. About midway on the larger wing feathers each is marked with a white spot, which spots form together a noticeable *white bar* on this part of the wing. The tail feathers are barred with buff, and all but the middle two have a broad white bar near their ends. There is a conspicuous white band across the lower part of the throat. The breast and upper throat are dusky or black. The remainder of the lower parts are barred transversely black and white, the latter often having a buffy tinge. The female has no white on the tail, and the throat patch is more restricted and buff in color.

The eggs are laid on the ground without an attempt at nest building. Often these are deposited in a gravelly field or on a bare rock, and not infrequently the flat roof of a house or other building, particularly such as have asphalt roofing covered with gravel, is utilized by the birds.

The eggs match their environment, being dull grayish or white, profusely marked with varying shades of grayish brown.

The birds range throughout Eastern North America north to Labrador, and breed from north of the Gulf States to that region. They winter in South America.

A much more gregarious bird than the Whip-poor-will, the Nighthawk may be seen both in spring and fall in the broad light of day migrating leisurely in bands of varying size, frequently many hundreds being associated together. They do not fly very low, but seek their insect prey at varying altitudes, generally above twenty and often several hundred feet from the ground. The well-known habit, during the breeding season, of diving from a very considerable height at great speed, and turning just before reaching the earth with a curious, sonorous, rushing sound, is very characteristic, and has been too often described to need further comment here.

Florida Nighthawk.
Chordeiles virginianus chapmani (Coues).

The geographical race of the Nighthawk breeding in Florida and on the Gulf Coast is smaller than its more northern prototype, being little more than eight inches long. It is also marked on the upper parts more profusely with white and light buff.

Western Nighthawk.
Chordeiles virginianus henryi (Cass.).

The Western Nighthawk is about the same size as its Eastern congener, but is grayer and paler in general tone. It is the representative of our Nighthawk throughout the Western United States, except on the treeless portion of the Great Plains, and a geographical race or form of that bird. The bird has been recorded from Northeastern Illinois.

The Red-headed Woodpecker is about nine inches and a half long. Adult birds have a hood over the head, neck, and breast of deep red,

Red-headed Woodpecker.

Melanerpes erythrocephalus (Linn.).

almost crimson. The back and the major portions of each wing are bluish black. A large spot on each wing, formed by the terminal half of its smaller feathers, and the rump and feathers above the tail are white. The tail is black, each feather having more or less white on its tips and edges. The chest, belly, and feathers below the tail are white. The belly is often tinged with reddish.

Immature birds present a similar general pattern but are colored quite differently. The red hood is replaced by a mottled one of dusky brown, sometimes having a red feather here and there. The blue of the back is not so rich as in the adults and is barred with ashy. The wings are more dusky like the tail, and the white area on them is replaced by feathers, marked or barred irregularly with black and white. The chest and belly are grayish white streaked more or less, especially on the sides, with dusky.

The birds nest in the characteristic woodpecker way, generally choosing dead trees in which to excavate holes to breed in. Four white eggs are laid about an inch long and three quarters of an inch broad.

The Red-headed Woodpecker is distributed through Eastern North America west to the Rocky Mountains. They breed on the coast locally from Florida to Northern New York and in the interior as far north as Manitoba. They winter regularly from Southern New Jersey southward and occasionally are found farther north in that season.

The Red-headed Woodpeckers are decided features in the bird world. They feed on insect larvæ as do most woodpeckers but also catch flying insects much after the fashion of flycatchers. They subsist on beechnuts in the coldest winter weather and are almost as fond of fruits and berries in the autumn as are their larger relative the Flicker. They are active cheerful birds at all times and much given to uttering their prolonged rolling call notes during their work.

On the prairies of Florida, in Polk, De Soto, and Manatee Counties, there is found a race of Owls which live in burrows in the ground, where

Florida Burrowing Owl.

Speotyto cunicularia floridana (Ridgw.).

they breed. They are the prototypes and close relatives of the Owls of the Prairie Dog Towns of the interior of North America. They are locally common, sometimes living in communities of considerable size and extent, and again in isolated pairs. They excavate their own burrows, and their chief food is small mammals, reptiles, and insects. They are diurnal birds, and may be found generally at all times of the day, perched on the little mounds of earth at the mouth of their burrows. When disturbed they often retire to their holes, but as frequently fly off a short distance and light on the ground.

The birds are about nine inches long, and of a grayish brown body color above, spotted and brokenly barred with white. Below white preponderates, especially on the throat, which is generally unspotted. The rest of the lower parts are more or less spotted and barred with grayish brown. Their long legs, *which are nearly unfeathered*, round heads, and yellow eyes will, with their size, readily identify them.

They are resident and local birds breeding in their burrows in April, at which season from five to seven white eggs are laid. These are nearly an inch and a quarter in length, and over an inch in their smaller diameter.

The Burrowing Owl of the Interior and West ranges from a little north of the United States south through parts of Mexico to Guatemala. It
Burrowing Owl.
Speotyto cunicularia hypogæa (Bonap.).
is a little larger than the Florida bird, more buffy in general tone, especially below. Otherwise it is similar. It has been recorded as accidental from Massachusetts.

The Hawk Owl is another diurnal owl partaking of many of the characteristics of the smaller Hawks. It is of extreme northern distribution,
American Hawk Owl.
Surnia ulula caparoch
(Müll.).
breeding from Newfoundland northward and is somewhat nomadic in its winter wanderings. It has been recorded casually at various points in the Eastern United States, as far south as Pennsylvania.

It is a bird about fifteen inches in length, has a round head and yellow eyes. Its general tone above is dusky. The head and neck are variegated with white spots. The back is barred with the same color. The tail is long, rounded and obscurely barred with whitish. The face is grayish white, and the regions about the ears dusky. The lower parts have a general ground color of grayish white, barred from the lower breast regularly with defined dusky markings. There is a dusky spot on the throat. The legs and feet are heavily feathered.

The birds nest in pine or other evergreen trees and sometimes in holes like the Screech Owl. The eggs are white, from three to six in number, and are a little more than an inch and a half long by almost an inch and a quarter in their smaller diameter.

The Snowy Owl scarcely needs description, so familiar has it become through many stuffed specimens. It is a bird about two feet in length, varying from almost uniform unspotted white to white quite **Snowy Owl.** heavily barred, and spotted with a dusky shade almost Nyctea nyctea (Linn.). black. The head is round and the eyes are yellow. The feet and legs are thickly feathered. The female is generally more barred than the male.

The nest is built on the ground or on a ledge on some rocky cliff. Very little material is used to construct a resting place for the eggs and young, usually some moss and feathers. From three to ten eggs are laid. They are white in color, and about two inches and a quarter long by an inch and three quarters in their other diameter.

The Snowy Owl inhabits the northern parts of the Northern Hemisphere. In North America they breed from Labrador northward, and occur in the more northern United States regularly in winter, straggling sometimes far south. There are records from South Carolina, Texas, and Bermuda.

The birds are diurnal and crepuscular in their habits, and are not night hunters. In Minnesota they prey largely on the Prairie Chickens and Grouse, but they feed also on mice and small animals.

The Short-eared Owl is a bird about sixteen inches long, with a general ground color varying from deep tawny buff to a very light shade of that color. This ground color is heavily streaked above with **Short-eared Owl.** dusky brown, which preponderates in this region. On the Asio accipitrinus (Pall.). upper breast the two colors are about evenly represented, but gradually the streaks of dusky become narrow, till they are often wanting on the belly, and are lacking on the feathers below the tail. The ear-tufts are short, inconspicuous, and *near together above the eyes.* The eyes are yellow or orange varying in individuals. The tail and the larger feathers of the wing are barred broadly with tawny buff and dusky brown.

The nest is built on the ground in open places, often in low meadows or salt marshes, among bushes or in tall grass. The birds lay from four to

284   SHORT-EARED OWL

seven white eggs, about an inch and three fifths long by an inch and a quarter broad.

They are found throughout North America, in fact are almost cosmopolitan. They breed irregularly and locally from Virginia and Kansas northward.

During the winter months these birds are often gregarious, and hunt by day as well as in the twilight. Flocks of from twenty to two hundred have been recorded, and I have personally seen such congregations in the winter in New Jersey, where I have also found them breeding near the coast. They are largely mouse hunters, though they occasionally prey on small birds.

In Southern Florida and on the Mexican border of the United States the Caracara is a common bird, and has in many localities the same scavenger **Audubon's Cara-** habits that characterize the Buzzards. With these birds **cara.** they frequent the vicinity of towns and villages, watching Polyborus cheriway (Jacq.). for any carrion, and lingering about slaughter houses for the offal thrown aside by the butcher. They also hunt live food for themselves, which consists largely of frogs and other small reptiles captured as the birds stalk about in the open country on the ground. When in South Florida, in the region about Lake Okeechobee, these birds frequently visited our camp, and were almost as familiar and tame, though not nearly as abundant, as the Buzzards. Their flight is essentially hawk-like, direct, swift, and graceful.

They were breeding during my stay in the above-mentioned region, and on April 13th I found a nest containing young half grown. This nest was built in a pine tree about thirty feet from the ground, and closely resembled an ordinary crow's nest in size and shape, being compact and small for a bird so large as a Caracara. Two other nests found the same day also contained young, and were about twenty feet from the ground in palmetto trees. Two eggs are generally laid. They are cream color or buff, strongly marked and washed with shades of dark brown. They are about two inches and one third long by rather more than an inch and four fifths broad.

The Caracara is a large bird, about two feet long, with a decidedly eagle-like head, long legs, and rather slim body. The face is bare. The crown is black with elongated feathers, forming almost a crest. The wings and lower back and the belly are black. The throat is yellowish buff. The region between the wings, the back of the neck, and the breast are yellowish buff finely barred with black. The tail is white, tipped and barred clearly with black. The sexes are alike, and young birds are duller, the buff areas replaced by dull browns generally unbarred.

Bird Studies.

The adult male Sparrow Hawk is about ten inches long. The top of the head is dull blue, and there is usually a pronounced crown patch of red-

American Sparrow Hawk.
Falco sparverius Linn.

dish brown. The back is reddish brown, more or less brokenly barred or spotted with black. The tail is red- dish brown, with a clear, broad black band near the end and next to a narrow white band forming the tip. The shoulders are dull blue spotted with black. The larger wing feathers are dusky, barred on their inner webs with white. There is a clear black bar under the eye, and another on the side of the head, defining the white region about the ears. There are frequently two black marks in the brown on the back of the neck. The lower parts vary from buff to white and the sides and flanks are more or less spotted with black. The female has the red of the back extending to the shoulders barred evenly throughout with dusky. The under parts are dull white streaked with dusky. The head is similar to that of the male but duller. The young show the difference in plumage, correlating with sex, that is emphasized in adult birds. The birds nest in deserted Woodpeckers' holes, and in hollows in trees. The eggs vary from three to seven in number and in color from cream to reddish brown. They are sometimes unmarked but generally are evenly specked with darker shades of their ground color. They are about an inch and two fifths long by an inch and an eighth broad. The birds range from Florida to Hudson's Bay and breed throughout the entire region. They winter from Northern New Jersey southward. Their chief food is insects, mice, and the smaller reptiles and they occasionally kill small birds, but are on the whole of great benefit to the farmer and fruitgrower.

The Cuban Sparrow Hawk has been recorded from the Florida Keys. It is a bird with the upper parts entirely dull slaty blue, except for some

Cuban Sparrow Hawk.
Falco dominicensis Gmel.

dark markings on the shoulders and back of the neck. The lower rump, the feathers above the tail, and the tail are bright reddish brown. The lower parts are white in some individuals and in others reddish brown. It is about the size of or a little smaller than the Sparrow Hawk.

Merlin.
Falco regulus Pall.

The Merlin of Europe is a bird about twelve inches long, grayish blue above and tawny yellowish with streak-

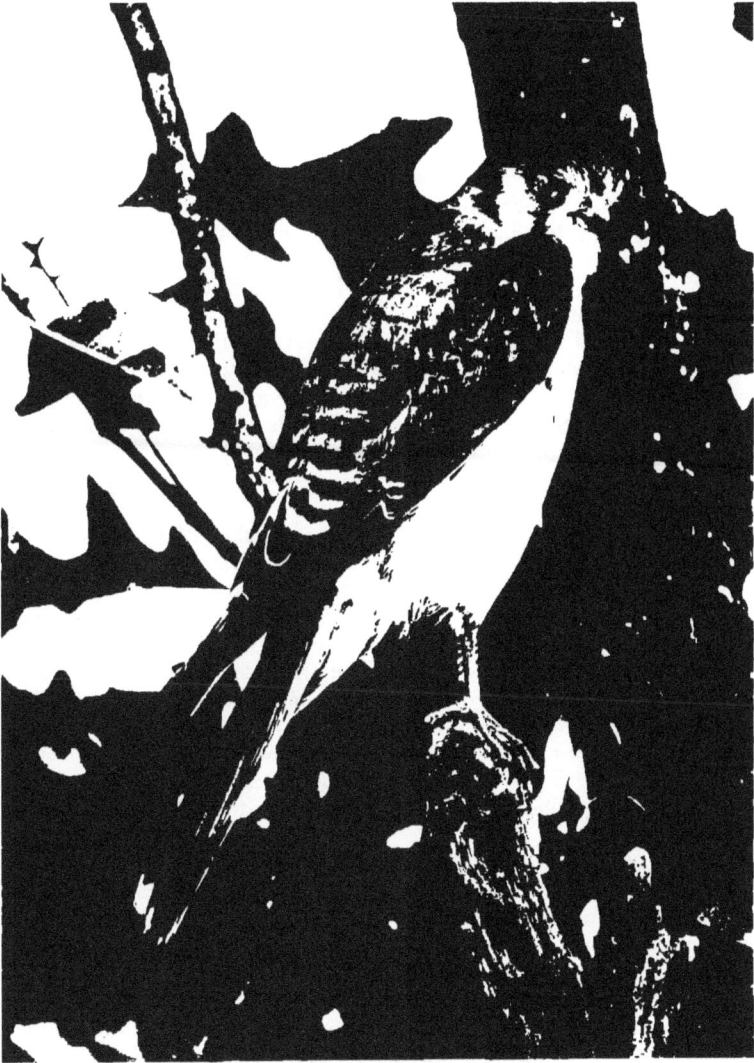

271   SPARROW HAWK.

ings of dusky below. The tail is like the back but barred with dusky or black. It has been recorded from Greenland.

The Kestrel is also a European bird that has been recorded once from Massachusetts, and so obtains a place in our fauna. It is a bird about four-teen inches long. The male has the upper parts, except

Kestrel.
Falco tinnunculus Linn.

the back and shoulders, leaden gray, which color also prevails on the tail. There is a black bar next to the ter-minal white one on the tail. The back and shoulders are reddish brown, or cinnamon, spotted with dusky. The forehead is buffy white. The under parts vary, from whitish to buff, and are streaked with dusky, on the breast, sides, and flanks. The female is reddish brown or cinnamon throughout above, including the tail, and barred or streaked with dusky. Below she resembles the male in coloring.

This is a Western species, appearing occasionally as far east as Illinois. The adult male is a bird about seventeen inches and a half long, and the fe-male is usually at least an inch and a half longer. The

Prairie Falcon.
Falco mexicanus Schleg.

birds are light brown above with a strong grayish tinge and obscurely but broadly barred with paler ash brown or bluish gray. The lower parts are white, with the sides and flanks streaked or spotted with dusky brown. These birds are finely formed, in the true falcon mould, rather slim but with obviously great power in the breast and wings. They are characteristic birds of the Great Plains and remind one much of the Peregrine Falcons in their flight and methods of capturing their prey, which consists of the smaller mammals and birds, even those as large as Prairie Chickens.

The birds nest both on the shelves of cliffs and in hollows of trees. They lay two or three eggs, which vary from cream white to light reddish buff in ground color, sometimes only speckled but often heavily spotted with warm shades of deep brown. They are almost two inches and a tenth long and an inch and two thirds broad.

The Prairie Falcon is found from the eastern border of the Great Plains to the Pacific, and as far north as the northern boundary of the Western United States. It winters in Kansas, Colorado, and southward.

There are two distinct kinds of Gyrfalcons that are found in Eastern North America. One, a bird of the Arctic regions and which has occurred casually in winter in the State of Maine, is known as the White Gyrfalcon. The other, known as the Gray Gyrfalcon, is also an Arctic bird, and is found in winter regularly south to the Northern United States.

Besides there are recognized two races of this last bird, the first known as the Gyrfalcon, whose range is from Northern Labrador and the coast of Hudson's Bay to Alaska. This bird has been taken in Rhode Island.

The second geographical race of the Gray Gyrfalcon is known to occur on the coast of Labrador and is found in winter regularly in Canada and Maine, and has been recorded from Long Island.

To recapitulate, there are two distinct kinds of Gyrfalcons, and in addition two geographical races recognized as belonging to the bird fauna of Eastern North America.

These birds are all about the same size, the males being about twenty-two inches long, and the females somewhat larger.

They nest on the face of rocky cliffs or in trees and lay three or four eggs which vary much in color, from white spotted with reddish brown to pale reddish brown spotted with darker shades of a similar color. The eggs are rather more than two inches and a quarter long, by about an inch and three quarters broad.

The color of the White Gyrfalcon is indicated by its name. The prevailing tone is *white.* This is usually narrowly streaked with black on the top of the head and neck. The remainder of the upper **White Gyrfalcon.** parts are transversely spotted with brownish gray. The
Falco islandus Brünn. lower parts and the tail are white. Frequently the tail shows traces of dusky barring on its central feathers and there are often obscure streaks of grayish brown on the white of the under parts. The feathers below the tail are *always immaculate white.*

The Gray Gyrfalcon *has the feathers below the tail* more or less *marked with dusky.* The upper are pale gray, the feathers barred or edged with white or buffy white. The tail is barred narrowly with sim-
Gray Gyrfalcon. ilar colors. The under parts are white streaked with
Falco rusticolus Linn. dusky markings. The flanks and legs are generally barred or spotted with brownish slate.

This is a darker bird in general color than the Gray Gyrfalcon. The dusky coloring prevails on the top of the head. The back is usually brownish gray or slaty without definite barring, or obscurely marked. The tail is colored like the back with narrow and indistinct barring of bluish gray. The under parts are streaked whitish and dusky grayish, particularly on the sides and flanks.

Gyrfalcon.
Falco rusticolus gyrfalco
(Linn.).

The Black Gyrfalcon has upper parts darker even than in its congener the Gyrfalcon, and uniform or unbarred. The tail sometimes shows broken or obscure barring. *The lower parts* are much *the same* as the *upper parts in tone and color.*

Black Gyrfalcon.
Falco rusticolus obsoletus
(Gmel.).

The Golden Eagle is found throughout the northern regions of the Northern Hemisphere. In North America it occurs at large, extending southward to Mexico. It is uncommon except in the more unsettled mountain regions. I have seen the birds in the vicinity of Asheville, North Carolina, on a few occasions, and they are frequent throughout the great mountain ranges of the West.

Golden Eagle.
Aquila chrysaëtos (Linn.).

The *feathering on the lower part of the leg extends to the base of the toes.* The general color of the plumage is dusky brown. This is relieved by the lighter yellowish brown color of the feathers of the head and neck, which are pointed or lanceolate in shape. The tail is *white* on the *basal half or two thirds,* and the remainder is blackish.

The eyes are light brown. The male bird is about two feet and six inches long, and the female is some six inches longer.

The Golden Eagle breeds on ledges or shelves of broken cliffs in remote and generally inaccessible places. The eggs, two in number, vary from white, almost immaculate, to dull white so heavily marked with varying shades of brown as to be almost obscured.

The birds, as I have observed them in Arizona, live mainly on the smaller animals and larger ground birds of the region, but they also are carrion feeders, and the carcass of a dead steer or horse is almost sure to attract them.

This is one of the large, heavily built hawks that are characteristic of field and meadow, where, perched like sentinels, they watch for the mice or other small animals that form their staple diet.

Ferruginous Rough-Leg.
Archibuteo ferrugineus (Licht.).

The Squirrel Hawk is a Western species, ranging east to Illinois. It is found as far north as the Saskatchewan River, and southward into Mexico. It breeds regularly from Kansas, Nebraska, and Colorado northward, and leaves the more northern parts of its habitat during the winter.

The birds are about two feet long. The feathering of the legs extends to the *base of the toes.*

The general color of the upper parts is rusty brown, the centre of the feathers being dusky. The thighs are bright rusty brown barred transversely with dusky. The tail is whitish, washed with rusty brown. The under parts are white, often with rusty brown barring on the sides and belly.

The birds nest rather low in trees or on shelves on cliffs. The nest is much like that of the Red-tailed Hawk in appearance, and built of similar material. Three or four eggs are laid. They are white or cream in color, marked with varying amount and differing shades of brown. The eggs are about two inches and a half long and an inch and nine tenths broad.

The Rough-legged Hawk is a bird similar in build to the Squirrel Hawk, but rather smaller, averaging about twenty-two inches in length. It

American Rough-legged Hawk.
Archibuteo lagopus sancti-johannis (Gmel.).

is found throughout the northern portions of North America, and breeding north of the United States, winters in the eastern regions as far south as Virginia.

The feathering on the legs extends to *the base of the toes.* In color these birds are dusky brown above. The feathers of the head and neck are broadly margined with whitish gray or light buff, and on the back this edging is narrower and not so light. The tail is *white* or *cream color* on its basal half. The terminal half has two or more grayish or dusky bars. The under parts vary from white to cream color, streaked and barred all over with dusky markings, often so concentrated as to form a dusky band across the chest.

From this phase of color there is a very wide variation toward a distinctly darker plumage, culminating in individuals which are almost or quite black, except on the forehead and the basal part of the tail.

The nesting sites are on trees or on shelves on cliffs. The eggs vary from three to five in number. They are white in color, frequently unmarked, but oftener washed or spotted with varying shades of brown. They are about two inches and a fifth long and an inch and three quarters broad.

ROUGH-LEGGED HAWK. FEATHERING ON FOOT.

The Mexican Goshawk is about seventeen inches long, and robustly built. Above the birds are deep bluish ash color, the top of the head and the back of the neck showing blackish or dusky streaks along the **Mexican Goshawk.** middle of each feather. The tail is black, with a narrow Asturina plagiata Schlegel. white tip, and barred by two or more bands of the same color. The feathers covering the base of the tail above and below are white. The rest of the under parts are white barred with deep ash. The birds nest on high trees, building much after the manner of Cooper's Hawk. The eggs are two or three in number, white in color, very faintly marked with pale reddish brown. They are about two inches long and an inch and three fifths in their other diameter.

The birds are southern in their distribution, ranging from Panama through Mexico and into our southern border, Arizona, etc. There is a single record of their occurrence in Southern Illinois.

Swainson's Hawk is another of the stout, heavy, " mouse " hunting birds of medium size, being about twenty inches long. It is a common bird of

Western America, being found regularly as far east as the Mississippi River, and there are numerous records of stragglers eastward to the Atlantic **Swainson's Hawk.** States. They breed regularly from Texas northward to

Buteo swainsoni Bonap. the Arctic regions and winter in the more southern parts of their North American range. The adult birds have the forehead narrowly white. The feathers of the upper parts are grayish brown, margined with buff or rusty brown. The tail is lighter than the back and is crossed by a varying number of dusky bands, which show more plainly below. The throat is white and there is a *broad cinnamon red area forming a band across the breast and chest.* The rest of the lower parts are white or cream color, more or less barred and spotted or streaked with shades of brown from dusky to pale reddish. From this phase of plumage the birds grade in every degree to a uniform dusky brown plumage known as the dark phase.

Immature birds resemble the adults, but lack the chest band of cinnamon red, the entire under parts being cream or buff in color streaked and spotted with dusky brown.

The birds nest in trees and in the giant cactus from ten to eighty feet from the ground. The eggs are colored from bluish white to cream, are frequently immaculate, but have more often markings of different shades of brown, which *vary* from light speckling to dark washes. They are two inches and a quarter long and about an inch and three quarters in their other diameter.

In the Middle Atlantic States our most common large hawk is the Redtailed or Hen Hawk. It is essentially a bird of the more open country, and **Red-tailed Hawk.** is often found away from the woods where some solitary

Buteo borealis (Gmel.). tree is a feature of an otherwise unbroken field or meadow. Late in the fall one of these birds will frequently select some piece of meadow or low marshy ground, where daily he may be observed presiding from his chosen tree over the destiny of the mice and other small animals that abound in such localities. Apparently in supreme repose, his eye is ever watchful and woe to the unlucky mouse that enters his range of vision. A sudden swoop, almost a fall from the perch, and the prey is secured, the bird is back on the limb, left a moment before, and the small victim is almost as quickly torn apart and swallowed. Such is the regular method of his life, and though now and again some poultry yard may be raided, this is not a Hen Hawk but the veriest cat among birds.

The Red-tailed Hawk is about twenty inches in length, robustly and heavily built. He may be known, when adult, by his bright bay tail, which generally has a narrow black band near the end and a narrow buff tip to each feather. His upper parts, the feathers of which are edged with rusty, buff, or buffy white, are grayish brown. The lower parts are buffy white, the upper breast is closely striped with dull reddish brown, the lower breast is much less and often not at all marked. The sides and chest are barred and streaked with dark dusky brown and the belly is generally immaculate. The feet are bare of feathers to the heel.

Immature birds are grayer above, and the tail is gray with many obscure dusky bars or bands crossing it. There is a more or less pronounced dusky area across the lower part of the chest.

The birds nest in trees well up from the ground, building a substantial structure of sticks, roots, and coarse material, lined with coarse grasses and the like.

Two to four eggs are laid, dull white in color, sometimes immaculate, but more frequently blotched or washed with shades of pale reddish brown. These birds are found throughout Eastern North America, as far north as the northern part of Hudson's Bay.

They breed generally throughout their range and winter south of the northern border of the United States.

Krider's Hawk is a near ally of the Red-tail and is similar in appearance. It has, however, a greater preponderance of white in its plumage, the head

Krider's Hawk.
Buteo borealis krideii
Hoopes.

frequently being very nearly white. The under parts are lightly, if at all, streaked and barred, and the pale red tail generally lacks the black bar near the tip.

These birds are found on the Great Plains from Texas to Minnesota, and have been recorded from Iowa and Northern Illinois.

Harlan's Hawk, another ally of our Common Red-tail, is found in the Gulf States and lower Mississippi Valley, ranging east to Georgia and Florida.

Harlan's Hawk.
Buteo borealis harlani (Aud.).

It is known to occur casually in Kansas, Iowa, Illinois, and Pennsylvania. It is a little smaller than the Red-tail Hawk, and in the old birds the upper parts are very dark as compared with that bird. The tail is a mottled mixture of rusty red, black, grayish, and white, either of which colors may predominate. It is generally banded narrowly near the end with black, and tipped with whitish. The lower parts vary from pure white to dusky or sooty brown.

This is the Western race of the common Red-tailed Hawk. The birds vary from a light phase of plumage, difficult to distinguish from the true Red-tailed Hawk, to a uniform dark or sooty brown, except

Western Red-tail.
Buteo borealis calurus
(Cass.).

the tail, which is like that of the ordinary Red-tail. There is every possible gradation between these two extremes. This form occupies the whole of Western North America, east to the Rocky Mountains, and south into Mexico. It has been recorded as a straggler in Illinois. In Arizona I have found these birds breeding at isolated points, on the desert in low mesquite trees, and in the giant cactus. Frequently the nests are so near the ground that a man of ordinary height may look into them.

The European Buzzard is about twenty inches long. In adult birds the upper parts are usually dark brown, mottled with a darker shade of the same color. The tail is grayish brown, marked with about

European Buzzard.
Buteo buteo (Linn.).

twelve transverse narrow bands of dusky brown or blackish. The eyes are *yellow*. The lower parts are dusky mottled brown on the breast and throat, becoming lighter and generally heavily barred on the belly, sides, and flanks.

It has much the habits of our common Red-tailed Hawk, but the prejudice existing in regard to that bird has obtained against this, even more strongly, and it is almost exterminated in England, where it was once a common and very useful bird. It is included here as an accidental straggler to North America, there being a record from the State of Michigan, which seems open to doubt.

The Mississippi Kite, or Hawk, is a rather small, compactly built bird, about fifteen inches long. In general, the tone of the adult birds is plumbeous or bluish slate color throughout, lightest on the

Mississippi Kite.
Ictinia mississippiensis
(Wils.).

head and neck. The wings and tail are dusky, the latter unmarked and the larger feathers of the former spotted with chestnut brown, more distinctly on their inner webs. The exposed parts of the smaller wing feathers are grayish, like the head and neck. The eyes are reddish orange. Immature birds have the head white, finely streaked with blackish. The back is dusky brown, each feather being tipped more or less with rusty brown or whitish. The lower parts are white or buffy white, often streaked with reddish brown. The larger feathers of the wings are dusky and unmarked. The tail is dusky, crossed by a varying number of grayish white bands, generally three in number.

These birds prey largely on insects and small reptiles, and much of their hunting is done on the wing. They are gregarious during their migrations or after the breeding season, when they are frequently observed in small parties. They breed in tall trees, building nests of sticks and twigs, lined with Spanish moss or dry leaves. Here two or three pale bluish white eggs are laid. They are usually unmarked, but often stained by the decaying moss and leaves lining the nest. They are about one inch and three fifths long, and one inch and three tenths in their smaller diameter.

The Mississippi Kite ranges east of the Rocky Mountains, as far north as South Carolina, on the coast, and to Kansas and Southern Illinois in the interior. It is a more abundant bird west than east of the Mississippi River, and breeds from Texas to Southern Illinois more commonly. It has been recorded casually as far north as Pennsylvania, Wisconsin, and Iowa. A few winter in the South Atlantic and Gulf States, but the great majority pass south to Central America.

This is not a common bird east of the Mississippi River, and though recorded from various localities from South Carolina and Florida to Southern

White-tailed Kite.
Elanus leucurus (Vieill.).

Illinois and Texas, it is probably more common in California than elsewhere in North America. Its habits are similar to those of the Mississippi Kite, and in its general economy it is not unlike that bird.

The adult birds are sixteen inches long and plain bluish gray above, becoming lighter or almost white on the head. There is a large patch of black in the region about the shoulders. The lower parts and tail are white. There is a black spot in front and partly around the eye. The eyes are red. The birds nest in trees, from thirty to fifty feet from the ground. The nest consists of an outer structure of sticks and twigs, lined with strips of bark and dry weeds and grasses. From three to five eggs are laid. Their ground color is creamy white, washed and marked, so as to be almost obliterated with

shades of brownish claret. They are about an inch and two thirds long by
an inch and one third broad.

The White-winged Dove is found on the southern boundary of the
United States, from Florida, where it must be regarded as rare, through

White-winged Dove.
Melopelia leucoptera (Linn.).

Texas to Lower California. It breeds in the United
States, in Southern Texas, New Mexico, and Arizona. It
is a pigeon, about a foot long, which has, in adult birds, a
conspicuous white patch on the closed wing. The tail is
slightly rounded, the middle feathers are brown, and the others slate or pearl
color, with broad white tips. The general color of the adult birds is grayish
drab, or brown, becoming slaty dove color on the rump. The head and
breast are warm dove color, and the sides of the neck show metallic lustre,
shading from green to purple. There is a spot of dark metallic blue just
below the ear. The remainder of the lower parts are pearl, shading into
white on the feathers below the tail. The female is duller, and immature
birds resemble the female.

The birds nest in bushes, cacti, or trees near the ground. They lay two
creamy white eggs, about an inch and an eighth long and nearly nine tenths
of an inch broad. This is one of the noticeable birds of Southern Arizona,
where they live on the dry deserts.

The Heath Hen, now restricted, so far as known, to the island of Martha's
Vineyard, Massachusetts, was, as late as the early part of this century, found

Heath Hen.
Tympanuchus cupido (Linn.).

in localities throughout the Middle States.
A bird of the open, wooded country, it differs essen-
tially in this characteristic from the Prairie Hen of the
West.

The causes that have led to its rapid extinction and present restriction
are obscure, for it is obvious that its use as a game bird, and consequent
persecution by gunners, cannot alone account for the existing results. There
can be no doubt that the settlement of a country has much farther reaching
results and influences on the original fauna than is generally ascribed to it.

We know well that many of the smaller song birds have notably increased in numbers with the clearing and cultivation of the country, that their
very habits and even life economy have undergone radical changes. The

Wood Thrush, no longer shy, though formerly a bird of solitude in the deep woods, now seeks the neighborhood of houses to breed and sings from the trees along village streets. A number of the swallows have notably changed their habits in so vital a matter as the location of their nests, utilizing the buildings that have been reared by man. I cannot imagine a much more radical change. The disappearance of the Wild Pigeon from the Eastern States in comparatively recent times, and the extinction of another bird of strong flight, the Labrador Duck, which was a common market bird in New York City seventy-five years ago, cannot be ascribed alone to the numbers killed by sportsmen. A combination of circumstances reaching much deeper than one or more apparent factors must be looked for to account for such results.

No doubt the insular situation of the survivors of the Heath Hen has been a factor in protecting them longer than their congeners of the mainland, but it is difficult to believe that in such wildernesses as still exist in New York and New England these birds have been exterminated by the gunner alone. The latest reports from Martha's Vineyard indicate the existence of at least several packs of Heath Hens, perhaps a hundred birds altogether, but they are not so numerous as they were ten years ago.

These birds are about the same size as and of similar appearance to the Prairie Hen. They are more broadly marked with buffy white on the feathers of the shoulders and the "pinnate" feather tufts of the neck are composed of not more than ten feathers each. These feathers are *noticeably pointed at their tips.* The female is smaller than the male and has the " pinnate " tufts of the neck much reduced, or rudimentary.

The eggs are laid in nests situated on the ground in oak woods at the base of a stump. They are olive buff in color, about an inch and three quarters long and a little over an inch and a quarter broad.

The Prairie Hen is about eighteen inches long. It is found on the prairies of the Mississippi Valley south to Louisiana and Texas, east to Ken-

Prairie Hen.   tucky and Michigan, and west through Kansas and Ne-
Tympanuchus americanus   braska to the eastern parts of North and South Dakota.
(Reich.).   It ranges north to Manitoba. It is generally resident, but migrates north and south in Minnesota, Iowa, and Missouri. The prevailing opinion seems to be that its eastward range is becoming contracted, and that it is extending the westward boundary of its range.

263   PRAIRIE HEN.

The prevailing color above is brown. The pattern is a barring of brown and black, spotted more or less with reddish brown. The "pinnate" tufts of the sides of the neck are each composed of more than ten narrow, stiff black feathers, crossed by reddish brown and buff marks, and *rounded off* at their tips. These feathers overlie a bare surface of skin, capable of being inflated. The rounded tail is short and almost covered by the body feathers above it; the outside feathers are a third shorter than the middle ones. The throat is buff, shading into the white ground color of the lower parts, which are barred with dusky brown or black. The female is similar to the male, a little smaller, and with the "pinnate" tufts of the neck much reduced, or rudimentary. The birds nest on the ground. They lay from ten to fourteen olive buff eggs, and occasionally these are spotted with reddish brown specks. They are an inch and seven tenths long and an inch and a quarter broad.

The Black Cock of Europe and Northern Central Asia has been introduced into Newfoundland, where it is said to have become naturalized. It is

Black Cock.
Tetrao tetrix Linn.

a large bird, about two feet in length, of a general glossy blue black, relieved by a white patch on the wing and by the white feathers under the tail. The four outermost feathers of each side of the tail are strongly curved outward, towards their rounded tips. The female is a smaller bird, of tawny brown color, barred and spotted with black. The outer tail feathers have little if any of the characteristics of those of the male bird.

The Prairie Sharp-tailed Grouse is one of the group of birds known as Prairie Chickens and sold annually in our markets with the Prairie Hen

Prairie Sharp-tailed Grouse.
Pediocætes phasianellus campestris Ridgw.

proper. It is rather shorter than that bird, being about seventeen inches long. The general color above is brownish buff, and this color marked and spotted with black closely matches the dry grass surroundings that prevail in the regions frequented by the birds. The exposed webs of the larger wing feathers are spotted with white and the shoulders are streaked with the same color. The middle tail feathers are colored much like the back and are about an inch longer than the other feathers of the tail.

The throat is buff, shading into whitish, which is the prevailing color of the lower parts. The breast is spotted with dusky or black arrow shaped marks, and the sides and flanks are barred and spotted with a like shade.

The female is smaller and has the middle tail feathers but little longer than the others.

The nest is built on the ground, and from ten to fourteen buffy eggs, usually dotted with fine reddish brown, are laid. They are about an inch and two thirds long and a little less than an inch and a quarter broad.

These birds inhabit the Plains east of the Rocky Mountains, are resident where they occur, and range as far east as Wisconsin and Illinois.

There are five kinds of Ptarmigans which are recognized by naturalists as occurring in Eastern North America.

In winter these birds are characterized by a general snow white plumage. In the breeding season a dark plumage is assumed that differs much in the several kinds.

In nesting the Ptarmigan usually gathers some coarse grasses and dry leaves, with which they line a shallow depression in the ground in a protected place. The number of the eggs is from seven or eight to as many as fifteen. The ground color varies from pale buff to deep brown, more or less specked and otherwise marked by differing shades of brown. In size they are about an inch and four fifths long, by an inch and a quarter in their other diameter.

These birds are feathered with a thick filamentous hairy-like covering on their feet that reaches to the ends of the toes, almost concealing the nails.

The Willow Ptarmigan is found throughout the northern part of the Northern Hemisphere. In America they migrate south in the winter to Sitka and the British Provinces. They have been re-
Willow Ptarmigan. corded as casual, or accidental, in New England. In
Lagopus lagopus (Linn.). summer the male, a bird about fifteen inches long, is reddish brown above, irregularly and finely barred, and mottled with black. The middle tail feathers are like the back. The remainder are blackish, tipped with white. The throat, breast, and sides are similar in color to the upper parts, and the belly is white. In the female the markings are more numerous and wider. In the winter the birds are entirely white, except the outer tail feathers, which are blackish with white tips.

Allen's Ptarmigan is only known to inhabit Newfoundland and we have

Allen's Ptarmigan.
Lagopus lagopus alleni
Stejn.

but little knowledge in regard to its general habits. Its summer plumage is unknown, but in winter it is a bird much resembling the Willow Ptarmigan save that the *shafts of the smaller feathers of the wing are black.*

The Rock Ptarmigan is a bird of Arctic America. It is found in Southern Labrador and southward to the Gulf of St. Lawrence, and in parts of Greenland.

Rock Ptarmigan.
Lagopus rupestris (Gmel.).

It is a rather smaller bird than the Willow Ptarmigan, being about thirteen inches and a half long. The general color of the males in summer is grayish brown, marked with a fine zigzag barring of black and buffy white. The feathers of the back are black in their centres, producing large blotches or irregular spots. The middle tail feathers are like the back, the others are blackish tipped with white. The lower parts are much the same in color and markings as the upper parts, but the belly is white. The female in summer is yellowish buff, spotted and barred irregularly with black. Below, the bars are more widely separated, the coloring resembling the upper parts.

In the winter these birds are white, except that the outer tail feathers are blackish tipped with white, and that the *region between the eye and the beak is black.*

Reinhardt's Ptarmigan is found in the northern part of Labrador and northward through Greenland. It is a bird closely resembling the Rock

Reinhardt's Ptarmigan.
Lagopus rupestris reinhardi
(Brehm).

Ptarmigan, but the character of the barring is broken and narrower. The female has the back and upper parts largely black, the feathers being bordered or spotted on their edges with pale buffy. The lower parts are light buff, barred broadly with dusky or blackish.

Welch's Ptarmigan is found in Newfoundland. It is a bird a little smaller than the Willow Ptarmigan. The male, in summer, has the upper parts dark grayish brown, spotted with black and finely mottled with black, dusky and grayish white, and buff. The tail is dusky gray brown and the

QUAIL OR BOB-WHITE. ADULT MALE.

middle feathers are *tipped with white.* The region in front of the eyes is black, and the head and neck are more coarsely barred than the back with similar

**Welch's
Ptarmigan.**
Lagopus welchi Brewst.

colors. The throat is white. The rest of the neck, breast, and sides much like the back. The belly is white. The female at this season is more broadly barred with black and grayish white mottled with buff. The central tail feathers are like the back and the others dusky or blackish. In winter the birds are white, with dusky tails which have the *centre feathers tipped with white,* and the *region in front of the eye is dusky or black.*

The Quail or Partridge, the Bob-white of our fields, is so well known as to need but a word in the way of description. The forehead, a line over

Bob-white.
Colinus virginianus (Linn.).

each eye, and the throat are white in the male bird, and these same parts are rich buff in the female. The upper parts are variegated with tawny chestnut, black gray, and buff. The lower parts are dusky just below the throat patch, then there is a mottled area of pinkish brown, black, and gray. The lower breast is indistinctly barred black and white on a tawny gray ground. This barring becomes more defined towards the belly. The sides and flanks are streaked with reddish brown, broken by buffy white spots with narrow black borders. The belly is grayish white, and the feathers below the tail reddish chestnut. The birds are about ten inches long.

The Quail is found in the Eastern United States, west to Eastern Minnesota. South they reach to middle Georgia, Alabama, Louisiana, and Texas. They range as far north as Southern Maine and are generally non-migratory. They breed on the ground, building rude nests in grassy places, and laying from ten to fifteen white eggs. These are about an inch and a fifth long by more than nine tenths of an inch broad.

The congener of our Bob-white in Florida is a much smaller bird, being about eight inches and a half long. The plumage is similar, but is much

Florida Bob-white.
Colinus virginianus floridanus (Coues).

darker throughout ; the areas of dark browns and blacks of the back being more extensive, and the rump grayer. The black, below the white throat patch, is often much emphasized, forming a conspicuous band across the breast. The barring of the chest is more evenly black and white, and the chestnut of the sides and flanks is deeper and occupies more of the area of the feathers, to the exclusion of the white spots with black margins.

The breeding and general habits are similar to those of the more northern bird, though the Florida Quail is equally at home in the pine woods of the region, as in " old fields " and grassy clearings.

This bird has been introduced into our fauna, and is thought to have become partly naturalized. It is a smaller bird than our Bob-white, about **The European Quail.** seven inches long, and has three longitudinal yellowish stripes on the head, is ash brown in general color, above Coturnix coturnix (Linn.). variegated, and striped both black and straw color. The lower parts are reddish brown, variegated with black, changing on the under neck to yellowish, and on the throat to whitish. These areas are more or less broken by bands or patches of dusky brownish.

IN MARSH AND SWAMP.

IN MARSH AND SWAMP.

D URING the warmer portions of the year, if you go to almost any marsh where flags, cat-tails, or reeds are growing, or on the salt marshes of the sea-shore, you will hear presently an unmistakable wren song, and very soon see the singer. The birds are extremely active, and their nervous energy finds vent in constant busy movement. They resent any intrusion or trespass on their haunts, and scold, as continuously as any Catbird or White-eyed Vireo under like circumstances. They are generally gregarious, even in the breeding season, many pairs nesting at no great distance from one another.

Long-billed Marsh Wren.
Cistothorus palustris (Wils.).

The nervous energy of the male bird after laying has begun seems to be redoubled, and it is no unusual thing to find him employed in building another nest at such times. Frequently several such structures are begun and carried to various stages of completion. Usually the outer wall is completed, and then another nest is begun. At one time in a small marsh, not more than forty feet long and some twenty-five wide, and only occupied by a single pair of birds, I found eight new nests. One of these contained five fresh eggs, and the others were to all appearance the result of the efforts of the male bird. I visited the place when it contained but two nests, and the others were built in the succeeding ten days. They were none of them ever used, save the one that contained five eggs, for breeding in.

The Long-billed Marsh Wren is about five inches and a quarter in length. The upper parts are deep olive brown on the head, often almost black on the sides, darker brown on the back, generally nearly black or very dark umber, shading into reddish brown, on the rump.

There is a clear white stripe over the eye, and the dark brown of the back is relieved by numerous fine clear white streaks. The wings and tail are barred dusky and lighter brown. The under parts are white or grayish white, washed on the sides and flank with pale brown.

The nest is a spherical structure, woven in the tall reeds and grasses which the birds frequent, with a small hole in the side for an entrance. It is firmly attached to the reeds which sustain it, and when new closely matches its environment in color. It is lined with fine grasses and plant fibres. From five to eight eggs are laid. They are usually a uniform rich brown in color; sometimes a white ground just shows through a profuse dotting of dark brown. Frequently the brown is relieved with numerous dots, specks, and marks of a deeper brown. The eggs are about two thirds of an inch long and a scant half-inch broad.

The Long-billed Marsh Wren is found in Eastern North America, north to Massachusetts and Southern Ontario. It winters from the North Gulf States, and in some localities farther south, to Eastern Mexico. It breeds from north of the South Atlantic and Gulf States through its North American range.

The coast region of South Carolina and Georgia is the home of a geographical race of the Long-billed Marsh Wren, known as Worthington's Marsh Wren. It is a rather smaller and paler bird above than the Long-billed Marsh Wren, the umber brown or black areas being restricted or wanting. It is grayer brown on the sides than its northern congener. Its general economy and breeding are like those of the Long-billed Marsh Wren.

Worthington's Marsh Wren.
Cistothorus palustris griseus Brewst.

Marian's Marsh Wren is a resident bird, similar in general character to the Long-billed Marsh Wren, on the Gulf Coast of Florida and probably westward to Louisiana. Compared with the Long-billed Marsh Wren it is somewhat smaller, and *the upper parts are* much darker in general color. The sides and flanks are about the same in tone as the rump. The rest of the under parts are grayish and the breast is generally, and the feathers below the tail *always*, barred with dusky.

Marian's Marsh Wren.
Cistothorus marianæ Scott.

The breeding habits are similar to those of the Long-billed Marsh Wren, spherical nests of grasses and reeds being suspended to the coarse grass in the salt marshes which these birds frequent. So far as I am aware the birds have not yet been found away from tide water.

The Short-billed Marsh Wren is about four inches long, and the character of its plumage above is a streaked mixture of white and buff and dark brown and black. The wings and tail are barred dusky brown or blackish and **Short-billed Marsh** grayish or reddish brown. The under parts are white or **Wren.** grayish white, shading into warm buff on the breast, sides, Cistothorus stellaris (Licht.). and flanks, and on the feathers below the tail.

The nest is placed on or near to the ground, and is spherical with a small hole on the side for an entrance. It is built of coarse grasses and reeds, and lined with finer grasses and plant down. It is generally concealed in the heavy tussock of grass which serves to sustain it. The eggs are pure white, generally immaculate, but sometimes marked with faint lavender and reddish brown specks. They are rather more than three fifths of an inch long and less than half an inch in their smaller diameter.

SHORT-BILLED MARSH WREN.

The Short-billed Marsh Wren does not always seek the wet marshes, but may be also found in damp meadows quite away from water. It is not so much a bird of the salt marshes as the Long-billed Marsh Wren, but is much like that bird in its general nervous habits and in its propensity to resent any intrusion upon what it considers its domain. During the breeding season its song is distinctly wren-like and very charming, and at all times of the year it informs you of its presence by guttural scolding notes similar to the Long-billed Marsh Wren and like some of the House Wren's tones. The birds are found in Eastern North America as far north as Southern New Hampshire, Southern Michigan, Southern Ontario, and Southern Manitoba. They range west to the Plains. They breed locally throughout their United States range, and winter in the South Atlantic and Gulf States.

The general color of the Maryland Yellow-throat is olive green above. This is relieved on the head by a black band across the forehead, extending **Maryland Yellow-** over the region immediately about the eye to the cheeks **throat.** and sides of the face. This broad band is bordered by a Geothlypis trichas (Linn.). narrow clear gray line behind. The throat is bright greenish yellow, shading into white on the belly and an olive brown on the sides and flanks. The adult female lacks the black band across the forehead and face, the whole top and sides of the head and face being olive green like the back, but often having a strong brownish tinge. Otherwise she is much like the male, but the yellow of the throat is paler and the white below more extensive. The adult male in the fall has the feathers of the black band tipped with clear gray, and is generally suffused to a greater or less degree with brownish, especially on the upper parts. Immature male birds have the black band more obscured by the gray tips of each feather, and sometimes barely discernible. They are more suffused with brown, and duller generally than adult males are in the fall. Fall females, old and young, are often very brown in general tone.

The Maryland Yellow-throat, while a swamp bird, preferring damp, wet places, yet generally selects such as have more or less bushes and thickets. In places like this you will meet him, or rather he will meet you, for of all the warblers he is only excelled by the Yellow-breasted Chat in curiosity and querulous scolding. The birds are about five inches long. The nest is built in low bushes and grasses close to or on the ground. It is made of dead leaves, grasses, strips of bark and plant fibre and with a lining of finer material. It is very bulky for so small a bird. From three to five eggs are laid. They are white, sparingly spotted or marked about the larger end with reddish and umber browns. They are seven tenths of an inch long and over half an inch in their other diameter.

These birds are found in Eastern North America, west to the Plains and north to Labrador and Manitoba. They breed from Southern Georgia northward and winter from the South Atlantic and Gulf States throughout the tropics.

The Florida Yellow-throat is closely allied to and is a geographical race of the Maryland Yellow-throat. It is a resident bird throughout Florida and **Florida Yellow-** Southern Georgia. It differs from the allied form in **throat.** having a proportionately longer bill and tail, in being Geothlypis trichas ignota Chapm. much browner on its upper parts, and in having a more in-

207   MARYLAND YELLOW-THROAT'S NEST AND EGGS.

tense and extended area of yellow below. The shading of the flanks and sides is darker and the black across the forehead and on the sides of the face is broader, with a wider edging of paler gray behind it.

It is a bird of the scrub palmetto, but also affects damp, swampy places, brackish bushy marshes, and the vicinity of cypress ponds. Its general habits and nesting economy are similar to those of its near relative.

This bird, which has been known to ornithologists for sixty-five years, having been described from specimens taken near Charleston, South Caro-

Swainson's War-
bler.
Helinaia swainsonii Aud.

lina, in 1832, has for a long time been lost sight of, and has only been found again in recent years. It is now known to be a rather common bird in portions of the South Atlantic and Gulf States, especially in the regions near where the original types were secured. It breeds locally, where it occurs throughout its United States range, and winters in the West Indies and Central America.

During three weeks in April which I spent at Dry Tortugas, Florida, the bird was recorded three times, one individual being found in the early morning, sitting on the bureau in the room where I slept. It had come in through the open window during the night or very early morning.

The birds are about five inches long, heavily built, and are a peculiar shade of olive brown above, including wings and tail. The top of the head is chestnut reddish brown, and there is a yellowish white stripe over the eye. The lower parts are obscure buffy white, grayer on the sides and flanks.

Swainson's Warbler is a bird of swampy thickets and cane-brakes and lives on or near the ground. It builds, in tangles of vines, near or over the water in such localities. The nest, usually ten to fifteen feet from the ground, is composed of leaves and lined with fine roots and pine needles. The eggs, three to four in number, are faint bluish white, three quarters of an inch long and nearly three fifths of an inch in their other diameter.

There is a bird that is very much like the Red-eyed Vireo, living for the greater part of the year in the salt water mangrove swamps of the Florida

Black-whiskered
Vireo.
Vireo calidris barbatulus (Cab.).

Keys. It is the Black-whiskered Vireo, so called from a dusky streak that extends from the corner of the mouth down the sides of the throat. The birds are about the same size as the Red-eyed Vireo and similar in color, but generally paler and duller. The crown is lead colored with a narrow black line defining it, and there is a noticeable light streak above the eye running along next to the black line described. There is an appreciable wash of yellow on the grayish white of the lower parts, especially on the sides and flanks.

On the Gulf Coast of Florida I found these birds quite commonly as far north as the mouth of the Anclote River. Even here they were migratory, coming late in April or early in May and remaining till October. Their song resembles that of the Red-eyed Vireo and their habits are similar. But so far as I am aware this is a maritime bird, and in Florida I did not find it away from the vicinity of the salt water.

The nest and eggs resemble those of the Red-eyed Vireo.

The Yellow-green Vireo is a bird of Mexico and Central America that has been recorded from Godbout, in the Province of Quebec, Canada. It is **Yellow-Green Vireo.** a rather larger bird than the Red-eyed Vireo, is much like that bird in the color above, but is greener. Below the Vireo flavoviridis (Cass.). *sides and flanks* are *bright olive yellow* and the feathers below the tail *sulphur yellow*.

The White-eyed Vireo is a bird of dense thickets and undergrowths, particularly such as border ponds and streams which have marshy borders. **White-eyed Vireo.** In localities of this sort he scolds any who dares to in-Vireo noveboracensis trude, and though often invisible to the eye his abrupt and (Gmel.). characteristic song ever makes you aware of the domain he has preëmpted.

The birds are small, about five inches and a quarter long, and their prevailing color seems at first glance to be a rather bright olive green. Examined more closely you find a bird with the entire upper parts of this shade washed more or less with gray. The wing and tail feathers are dusky, with the exposed edges of the feathers like the rest of the upper parts, becoming whitish or light yellow on some of the smaller feathers. There are two yellowish white bars on each wing. The region in front of the eyes and a ring about each are bright yellow. The throat and belly are white, and the breast and sides are greenish yellow shading into the white of these regions. Adult birds in spring and summer have *white eyes ;* in fall many, including *all the immature* birds, have *brown eyes.*

WHITE-EYED VIREOS.

The nesting is much like that of the Red-eyed Vireo, and the eggs are hardly to be distinguished from those of that bird. They are three quarters of an inch long and more than half an inch broad.

These birds are found in Eastern North America north to New Hampshire and Minnesota and west to the Rocky Mountains. They breed from Northern Florida and the Gulf States northward, and winter from Florida to Guatemala and Honduras.

The Key West Vireo is the geographical race and closely related ally of the White-eyed Vireo, found in South Florida. Here these birds are resi-

Key West Vireo.
Vireo noveboracensis maynardi Brewst.

dent and do not migrate. They are essentially like the White-eyed Vireos in their habits and general economy. A little shorter than the White-eyed Vireo, they are generally *grayer* or paler in color and less yellow below. The bill is larger and the birds generally are more robust. The color of the eyes is pearl gray in adults and brown in immature birds.

The Swamp Sparrow is a bird nearly six inches long, whose prevailing colors are deep brown above and pale gray below. During the warmer portions

Swamp Sparrow.
Melospiza georgiana (Lath.).

of the year the bird has a deep chestnut crown. The forehead is dusky or black. There is a grayish stripe above the eye and a black or dusky line behind it. The back of the neck is ashy gray, with sometimes a few dusky stripes. The reddish brown back is relieved by broad dusky or blackish streaks, each feather being bordered with pale buff, often ashy in tone. The throat is white, and the breast is grayish, shading into grayish brown on the sides and flanks and into white on the belly. In winter the birds are similar, but the crown is streaked with black and grayish on its chestnut ground.

The birds nest on the ground much like the Song Sparrow, and the eggs are very similar to those of that bird, but generally more heavily marked and washed. They are about three quarters of an inch long and rather less than three fifths of an inch broad.

The birds are seldom found away from thickets in marshes or damp meadows, and where the marshes are covered with "cat-tails" and other tall grasses and reeds they abound.

These birds are distributed over Eastern North America west to the

Great Plains. They range north to Newfoundland and Labrador. They breed from Pennsylvania and Northern Illinois north, and winter from Massachusetts and Southern Illinois to the South Atlantic and Gulf States.

The Seaside Sparrow or Finch is a maritime bird, confined almost exclusively to the salt or brackish marshes of the Atlantic seaboard. Here

Seaside Sparrow.
Ammodramus maritimus
(Wils.).

it is found from Massachusetts to Georgia in the breeding season, and it winters from the Virginia Coast to Georgia. The bird is about six inches long, is stoutly built, heavily feathered, and has a narrow tail.

In full plumage it may be distinguished by an area of yellow just in front of the eye, and the same color is noticeable on the bend of the wing. The upper parts are olive green with a decided gray tone. The wings and tail are browner. A dusky streak on either side defines the grayish white throat. Grayish white prevails on the lower parts. The breast is obscurely streaked with darker gray, and the sides and flanks are washed with the same color. In fall and winter there is a buffy wash on the breast.

SEASIDE SPARROW.

The birds build on the ground in a grass tussock or among reeds. The nest is made of coarse reeds and grasses, lined with finer grasses. Three or four eggs are laid. These are white, closely speckled with reddish brown most profusely at the larger end. They are about four fifths of an inch long and rather more than three fifths of an inch broad.

The Dusky Seaside Finch so far as known is found only in the region about Titusville and Merritt's Island, on the Indian River, Florida. It is a bird about the size of the Seaside Sparrow. The same yellow areas obtain. The upper parts are black, each feather having a narrow edging of grayish olive green. The lower parts are white streaked all over evenly with black. The nest and eggs have never been obtained and but few collectors or naturalists have seen these birds alive.

Dusky Seaside Sparrow.
Ammodramus nigrescens Ridgw.

Scott's Seaside Finch is the representative, resident, geographical race allied to the more northern Seaside Finch found on the Alantic Coast from South Carolina to Northern Florida, and on the Gulf Coast of Florida north of Cape Romano. It is similar to its northern ally in form, but a little larger and much darker in general color. Above the prevailing tone is dusky brownish olive. The under parts are darker grayish white, and the breast, sides, and flanks are streaked distinctly with dusky or blackish.

Scott's Seaside Sparrow.
Ammodramus maritimus peninsulæ Allen.

The general habits of this bird are similar to the Seaside Finch proper and their breeding economy is almost identical.

The Louisiana Seaside Sparrow is much like Scott's Seaside Sparrow, but is still darker. The dusky streaks on the back are broader and blacker and except on the middle of the back are bordered with broad streaks of pale ash gray. The top of the head and back of the neck are strongly tinged with brown. The regions back of the ears and the chest are strongly tinged with buff, which on the chest is streaked with dusky or blackish.

Louisiana Seaside Sparrow.
Ammodramus maritimus macgillivrayi Aud.

This bird is found in the coast region of Louisiana and extends its migrations to the coast region of Texas.

The Sharp-tailed Sparrow or Finch is another maritime bird, confined almost exclusively to the salt and brackish marshes of the Atlantic sea-coast. They breed from the coast of New Hampshire to South Carolina and winter on the coast from North Carolina to both coasts of Florida. The salient feature of this bird is a rather short tail with narrow pointed

Sharp-tailed Sparrow.
Ammodramus caudacutus (Gmel.).

feathers, the outer ones being much the shorter. The birds are about five inches and four fifths long. The prevailing color of the upper part is olive green with a decided brown tone. The crown is olive brown, divided by a grayish median line. The region about the ears is gray. There is a buff stripe above the eye defining the crown, and a buff stripe on either side of the throat, reaching to and defining the gray region about the ears. The bend of the wing is yellow. Each feather on the back is edged with grayish or obscure white. The lower parts are generally white, washed on the breast and sides with buff, *which regions are definitely streaked with black.*

SHARP-TAILED SPARROW.

The birds build on the ground a nest of grasses, reeds, and seaweed, lined with fine grasses. Three or four eggs are laid. These are white finely dotted with reddish brown. They are a little more than three quarters of an inch long and nearly three fifths of an inch broad.

This form of the Sharp-tailed Finch is found during the breeding season in the marshes, generally near the salt water of Southern New Brunswick, Prince Edward's Island, and Nova Scotia. The birds are a little larger than the Sharp-tailed Finch and similar in color on the upper parts. The *throat* as well as the breast and sides are *pale buffy* and obscurely streaked with ash gray.

Acadian Sharp-tailed Sparrow.
Ammodramus caudacutus subvirgatus Dwight.

The Acadian Sharp-tailed Sparrow migrates south along the Atlantic Coast to South Carolina in winter, and during that season and in their

migrations they are often associated with both the Sharp-tailed Finch and
Nelson's Sharp-tailed Sparrow.

ACADIAN SHARP-TAILED SPARROW.

Nelson's Sharp-tailed Finch or Sparrow is rather smaller than the Sharp-
tailed Finch, being about five inches and a half long. The upper parts are
Nelson's Sparrow. rather darker and olive brown in color, and *the margins*
Ammodramus caudacutus of whitish to each feather *are broader.* *The lower parts,*
nelsoni Allen. except the whitish belly and feathers below the tail, *are*
deep buff with *little* or *no black streaking.*

These birds are the representatives of the Sharp-tailed Finch breeding
on the fresh water marshes of the interior from Northern Illinois north to
Dakota and Manitoba.

In their migrations they are found on the Atlantic seaboard from the
Carolinas to New England, and they winter on the Atlantic Coast and Gulf
Coast from South Carolina to Texas. During their migrations and in winter
they are often found associated with the Sharp-tailed Finch.

In the spring and fall migrations Rusty Blackbirds are found in pairs or
small companies in meadows and such swamps as are more or less covered with
Rusty Blackbird. low growths of bushes. Now and then I have found them
Scolecophagus carolinus feeding, in dryer places, on ripe dogwood berries, but the
(Müll.). earth is their general feeding ground, where seeds, small
fruits, insects, and worms are their diet.

They are not so noisy or musical, and are less frequently seen because of their haunts than the Crow Blackbirds, and yet their presence is often betrayed by characteristic blackbird-like notes, and in the spring their efforts at song are very agreeable.

The birds are rather more than nine inches and a half long. *Their eyes are yellow* or *straw color*. In the breeding season the males are unbroken glossy bluish black. The female is at the same period of a general deep dull lead color shading in the upper parts into dull greenish having an olive gloss. These colors prevail in the sexes at other seasons of the year but are suffused or concealed more or less, varying greatly in individuals, by rusty brown and shades of buff.

The birds are northern in their breeding range, choosing a variety of nesting sites from low evergreen trees to bushes and even the ground. The nests are built of twigs and grasses, and are lined with fine grasses. The birds lay from three to six or seven eggs. These are pale sage green in color, or greenish white, thickly marked with blotches of different shades of brown. They are about an inch in length and a little less than three quarters of an inch in width.

The birds are distributed over Eastern North America, west to the Plains. North of the United States they range west to Alaska. They breed from the northern borders of the United States northward. They winter from Virginia southward.

Brewer's Blackbird occurs in Western North America from the Great Plains to the Pacific Ocean. It ranges from the region of the Saskatche-

Brewer's Blackbird. wan south to the table-lands of Mexico, breeding in the
Scolecophagus cyanocepha- northern parts of its range, and at high altitudes in the
lus (Wagl.). mountains, and wintering in the more southern part of

its range. It is of casual occurrence in Illinois and Louisiana, and has been recorded once from South Carolina. The birds are somewhat larger than the Rusty Blackbirds, averaging about ten inches long.

The differences in color correlated with sex are similar, the males being clear black with a purplish undertone instead of blue, which is most noticeable on the head and neck. The female is brownish slate in color, and decidedly glossy olive green with a brown tone on the head and neck. In the changes of color that correlate with season, the obscuration of the ground color by rusty and buff is much less in these birds than in the Rusty Blackbird. The breeding habits and eggs of these birds are similar to those of the Rusty Blackbird.

The Red-winged Blackbird is one of the familiar birds seen in our springtime walks through low meadows or about marshy ground. The birds **Red-winged Black-** are about nine inches and a half long. The male when **bird.** fully adult is clear black throughout, except on the Agelaius phœniceus (Linn.). shoulders, which are the brightest scarlet, divided from the black parts of the wing by a band, varying from deep buff to almost white in color. Many individuals, presumably younger birds, during the breeding season have the black feathers edged or tipped to a greater or less extent by rusty, buff, or gray. All the males in the fall have the black feathers edged or tipped in a like manner, and in young birds of the year the black is often much obscured by rusty and buff.

RED-WINGED BLACKBIRD. ADULT MALE.

The female averages a little smaller than the male in size, and in summer is streaked above with dusky brown and black, rusty and buff. Below, this streaking is much more definitely black and gray, or black and white. The shoulders are generally obscurely tinged with deep crimson and the throat with deep orange or warm buff. In the winter all of this coloring is much suffused by rusty and buffy.

The nests are placed in low bushes or reeds, and are built of coarse grasses, weeds, and plant fibres, lined with a layer of fine soft grasses. The eggs vary from three to five in number. They are pale bluish, streaked and spotted and marked in zigzag lines with deep shades of brown. They are nearly an inch long and almost seven tenths of an inch broad.

These birds are distributed throughout Eastern North America to New Brunswick and Manitoba. They breed, except in South Florida and the Gulf Coast of Louisiana, throughout this area. They winter from Virginia southward.

The Red-winged Blackbirds which breed in the southern two thirds of the peninsula of Florida and on the Gulf Coast of Louisiana are smaller and have more slender bills than their northern congeners.

Florida Red-wing.
Agelaius phœniceus florida-nus Maynard.

The streaking of black on the lower parts of the female is not so broad, and the red on the shoulders is generally more noticeable. They are a local geographical, resident race of the Red-winged Blackbird, grading into that species. The breeding and general habits are very like those of the Red-wing.

The Yellow-headed Blackbird is about ten inches and a half long. The male has the head, neck, and chest yellow, which varies in tone from crimson to deep orange, sometimes being deep salmon pinkish.

Yellow-headed Blackbird.
Xanthocephalus xantho-cephalus (Bonap.).

There is an area of pure white on the shoulders. The remainder of the plumage is black. The region in front of the eye and the extreme upper throat is black or dusky. Old males in the winter have the yellow, especially of the top of the head, more or less obscured by the orange brown tips of each feather. The female is dusky brown, and has the throat and chest obscurely yellow, the feathers on the breast often being whitish. The female is somewhat smaller than the male. Immature birds resemble the female, but are generally darker.

The birds build generally in marshes, where the nests are attached to reeds. They are built of leaves, coarse grasses, and bits of marsh plants and lined with finer grasses. Four or five eggs are laid. These vary from white or grayish white to greenish white and are evenly specked with varying shades of brown. They are rather more than an inch long and nearly three quarters of an inch in their other diameter.

Yellow-headed Blackbirds are gregarious, even during the breeding season, at which times they are associated in small communities. During other parts of the year they are often congregated in enormous flocks, roosting in swamps and marshes and dispersing over the fields during the day, and often visiting the barnyards where cattle are kept. The birds are chiefly of Western North America, migrating as far north as Manitoba and south to the Valley of Mexico. They are found regularly east as far as Wisconsin and Illinois, and have been recorded as stragglers from Massachusetts, the District of Columbia, South Carolina, and Florida.

The Ivory-billed Woodpecker is the largest of North American Woodpeckers. It is a truly magnificent and remarkable bird, and is an inhabitant of the great wooded swamp areas of the more southern States. By no means as rare as many suppose, the distribution of the bird is quite local. A few years ago in Florida I saw eleven of these birds together, working on a piece of girdled timber near a cypress swamp. The trees were dead and sufficiently decayed to offer fine feeding ground for the larvæ of a kind of large boring beetle. This kind of food is much sought after by these giant Woodpeckers, and so fertile a hunting ground doubtless had attracted all of the birds living near. Their call notes, *kate, kate, kate,* were constantly uttered, and with their ceaseless hammering, and the flying of large detached chips, made a busy and noisy scene never to be forgotten.

Ivory-billed Woodpecker.

Campephilus principalis (Linn.).

Ivory-billed Woodpeckers are about twenty inches long, and their prevailing colors are black and white. They have a very conspicuous and pointed recurved crest on the head, which is bright scarlet in the adult male, and black, with sometimes an admixture of a few scarlet feathers on the forehead, in the female. A broad white stripe starting below each eye passes down the side of the neck and the two meet on the back. There is a large and conspicuous patch of white on each wing. The bill is white and the eyes are pale yellow.

This bird had formerly a much wider distribution than at present. It was found late in the last century and in the early part of the present one, as far north as North Carolina, and south to Texas on the coast, and in the interior up the Mississippi Valley to Missouri, Southern Illinois, and Southern Indiana. It is now restricted to the lower Mississippi Valley and the Gulf States, being probably most abundant in Western Florida, especially in the great cypress swamps, southwest of Lake Okeechobee, and in the vicinity of St. Marks, in the northwestern part of the State.

The birds nest in characteristic Woodpecker fashion, and lay white eggs. These are about an inch and two fifths long by about an inch in width.

On the 17th of March, 1887, I found near Tarpon Springs, Florida, a nest containing a single young one, presumably about ten days old. The hole for this nest was excavated in a cypress about forty feet from the ground, the entrance was oval, one diameter being about three inches and a half, the other four inches and a half. The hole was some fourteen inches deep, and had apparently been used for a breeding place before.

The Marsh Hawk is a long slimly built hawk, about twenty-one inches from the tip of the bill to the end of the tail. You will see the birds coursing over meadows and fields, particularly in the spring and fall, and a large white patch just above the roots of the tail will serve as a badge by which they may be known in any plumage, even at a long distance.

Marsh Hawk.
Circus hudsonius (Linn.).

The adult male is bluish gray above, except for the white spot on the rump ; the throat and upper breast are bluish gray. The rest of the lower parts are white, more or less barred and spotted with light reddish brown. The tail is gray barred indefinitely with dusky or blackish. The female is dusky above, the head, neck, and breast streaked with rusty brown. The shoulders are marked and spotted with a similar shade. The white rump spot is much the same as in the male. The tail feathers are barred with ash and dusky on the two middle ones, and with rusty buff and dusky brown or black on the remaining ones. The prevailing color below is rusty buff, streaked on the lower neck and breast, and much more finely and definitely striped on the sides, flanks, and belly with dusky and rusty brown. The eyes are yellow, varying from straw color to orange. Immature birds are darker and much more rusty above, and clearer almost unmarked rusty below. They, too, have the clear white rump patch.

311   HEAD OF MARSH HAWK. FEMALE.

The birds nest on the ground, generally in marshes or damp meadows, but sometimes in dry fields. From three to six dull bluish white, unmarked eggs are laid. They are about an inch and four fifths long and an inch and two fifths broad.

The birds are found in North America at large, and breed throughout their range. They occur from the South Atlantic and Gulf States south to the Isthmus of Panama.

BY STREAM AND POND.

BY STREAM AND POND.

THE Large-billed or Louisiana Water-Thrush is one of the most notable of the *land water birds.* You will find him about streams of varying size from the tiny wood brook or the rushing torrent to the placid flowing river. On such a stream as the last, the Kanawha River in West Virginia, my first acquaintance with this bird was established years ago, and

Louisiana Water-Thrush.

Seiurus motacilla (Vieill.).

at that point this Water-Thrush was very common. The river being then low there were very considerable gravelly beaches on either bank, and a rather small bird with many Sandpiper-like habits was conspicuous, walking or running along the ground, whose constant "tipping" motion seemed almost an imitation of the Spotted Sandpiper. Its food was of a similar kind to that of the Sandpiper, larvæ and worms and insects that frequented these gravelly shores. But when, satisfied for the time being, the bird mounted to some twig or branch and burst into a wild and peculiarly beautiful song, all thoughts of an affinity with Shore-birds were at once dispelled, and only wonder and admiration for his musical attainments remained.

This bird is about six inches and a third long. There is a clear white line over the eye and running well back on the sides of the head, the only mark relieving his otherwise unbroken coat of rich olive brown. Below, white predominates, being unmarked on the throat and belly, and streaked on the breast, sides, and flanks with dusky brown or black. There is a decided tinge of warm buff in the white of the flanks.

For a nesting place the birds seek some bank undermined by water, where roots of trees are thus exposed, or where some tree has been overthrown by the wind, forming similar conditions. Here a nest of leaves and fine twigs and roots is built, lined with grasses and plant fibres. From four to six white eggs are laid, spotted more or less distinctly with varying shades of brown. The eggs are about three quarters of an inch long, and nearly three fifths of an inch in their other diameter.

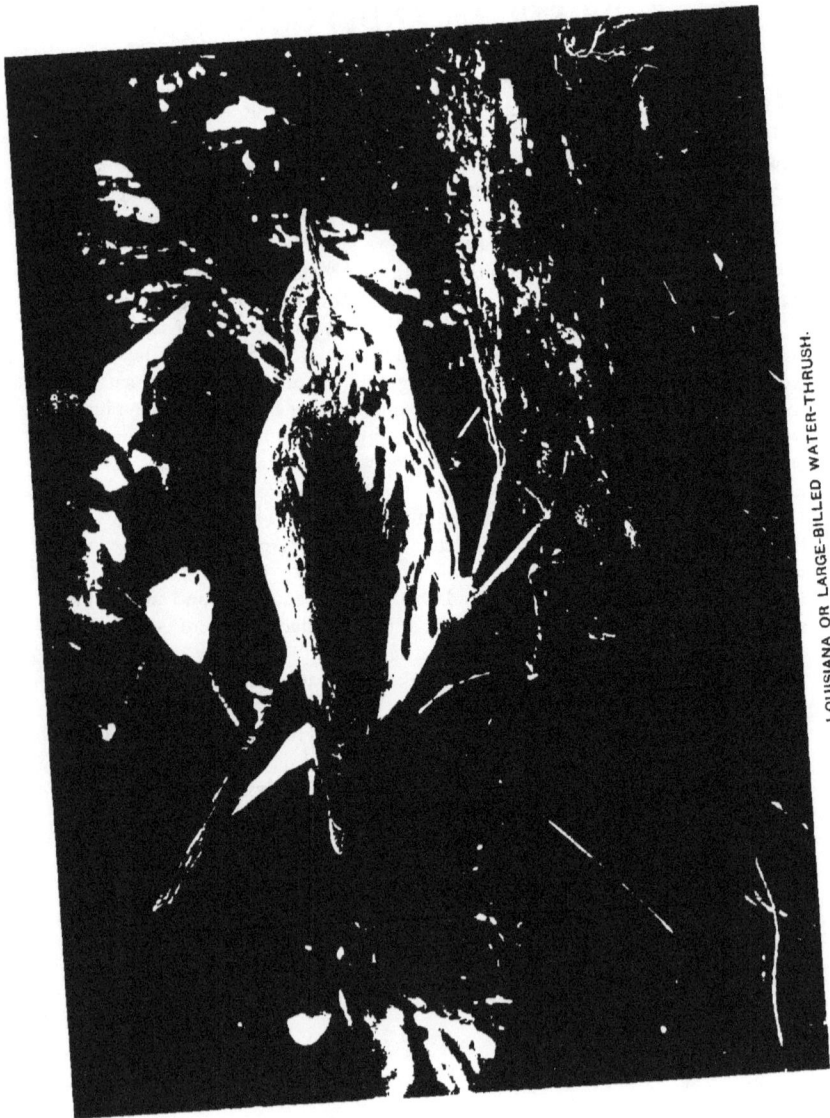

LOUISIANA OR LARGE-BILLED WATER-THRUSH.

316

The birds are found in the breeding season as far north as Southern New England and New York. In the interior they range farther north to Southern Michigan and Minnesota. They breed south at least as far as North Carolina, and winter in the West Indies and Central America.

This Water-Thrush passes through the Eastern United States west to Illinois during its migrations. It breeds from the Northern United States

Water-Thrush.
Seiurus noveboracensis
(Gmel.).

north to Arctic America. It winters from the Gulf States to Central America and Northern South America.

It is a smaller bird than the Large-billed Water-Thrush, about six inches long. The unbroken olive brown upper parts are relieved by a sulphur yellow or buffy line above the eye, extending well back on the sides of the head. The lower parts are pale sulphur yellow and are *streaked throughout* with dusky brown or blackish.

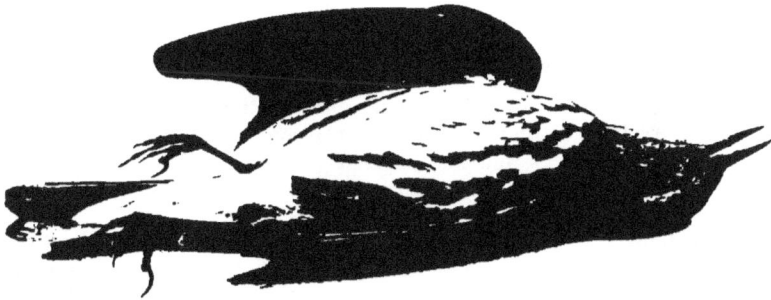

WATER-THRUSH.

Its nesting places are similar to those chosen by the Large-billed Water-Thrush. The nests are built much like those of that bird, save that mosses form a large factor in their construction. The birds lay from three to five white or creamy white eggs, marked with varying shades of brown, mainly at the larger end.   They are a very little smaller than the eggs of the Large-billed Water-Thrush.

The habits of the two birds are much alike, streams and rivers being congenial to both.   The more northern bird, however, seems the less shy of the two, and is often found during its migrations in damp places in woods well away from water.

Grinnell's Water-Thrush is a close ally and the Western geographical race of the common Water-Thrush just described.   It is a rather larger bird,

Grinnell's Water-Thrush.

Selurus noveboracensis notabllis (Ridgw.).

being about six inches and a quarter long.   It is darker above in tone, and lighter below, being often almost white.   The line over the eye is also clearer white.

Its habitat is from Illinois westward to California and north into British America.   During its migrations it is found regularly east-ward to Virginia and sometimes to New Jersey.   Its winter habitat is from the southern border of the United States, including the Gulf States, through Mexico and Lower California to Northern South America.

The Yellow-throated Warbler is a bird of the Southeastern United States and is more generally found in the vicinity of water.   The birds are

Yellow-throated Warbler.

Dendroica dominica (Linn.).

found regularly as far north as Southern Maryland and Virginia and have been recorded as accidental in New York and Massachusetts.   They breed throughout their United States range, and winter from Florida southward through the West Indies.

The birds are about five inches and a quarter long,   Their prevailing color above is bluish gray.   The forehead is blackish or dusky.   There is a pronounced white line over the eye becoming yellow *in front of it*.   There are two white wing bars, and the exposed edges of the dusky wing and tail feathers are gray.   The outer tail feathers have a patch of white near their ends.   The sides of the face and sides of the throat are black or dusky.   There is a white patch on each side of the neck.   The throat and breast are bright yellow, the belly is white, as are the sides and flanks, which are streaked with black.

The birds build a nest of strips of bark and various plant fibres, and where the Spanish moss grows in their breeding range it is utilized by the birds both as a part of the structure and sometimes as a location in which to place the nest itself. This is generally high up from the ground. Four or five white eggs, marked mainly at the larger end with reddish and olive shades of brown, are laid. They are about seven tenths of an inch long and a little more than half an inch wide. The birds find their insect food in the higher trees and frequent by preference those in the vicinity of water. They remind one somewhat in their motions of Titmice and again of Creepers, and are among the more musical of their allies.

The Sycamore Warbler is rather a smaller bird, with a relatively shorter bill, than the Yellow-throated Warbler, and is otherwise much like that bird except that the *region in front of the eye is white* instead **Sycamore Warbler.** of *yellow.* These birds are closely allied. The one under **Dendroica dominica albilora Ridgw.** consideration is distributed in the breeding season throughout the Mississippi Valley as far north as Southern Indiana and Illinois. It is found east, as far as the western edge of the Carolinas, and passes south in winter mainly through Texas to Mexico and Central America. It also winters more sparingly from Southern Florida and the Island of Jamaica southward. Like its congener it prefers high trees in the vicinity of water and is a notable musician among the Warblers. The breeding and nesting economy are similar in the two birds.

The Prothonotary Warbler is a bird of water-ways and streams and is Southern in its distribution. It breeds from the Gulf States north to Virginia, on the coast, and in the interior to Southern **Prothonotary Warbler.** Illinois. It winters in tropical America. **Protonotaria citrea (Bodd.).** A rather robust bird about five inches and a half long, the Prothonotary is conspicuous even among the Warblers for its beauty of plumage and elegance of carriage. The entire head and under parts are brilliant orange of a peculiar rich shade, becoming lighter on the belly and feathers below the tail. The back is olive gray with a yellowish tone changing to clear bluish gray on the rump. The wings and tail are nearly the color of the rump. The tail has much white on the inner webs of all but the middle feathers. The female is much like the male but paler and duller and whiter on the belly. Immature birds resemble the adult female.

The birds are generally to be found in low growths near water and not far above the ground. The nest is built in some hole or hollow in a tree or stump at no great height. Such a cavity is lined with fine twigs, roots, mosses, and strips of bark, with a finer lining added of plant down, fine grasses, and some feathers. In such a nest these birds lay from four to six white eggs, thickly speckled with rather large dots and markings of varying shades of reddish brown. The eggs are less than seven tenths of an inch long and upwards of half an inch broad. The song of these birds is loud and clear, and notable among Warblers.

The Boat-tailed Grackle is a Southern bird, generally aquatic in its habits, and is the *largest blackbird* to be found in the area being treated of. The

Boat-tailed Grackle.
Quiscalus major Vieill.

male is about sixteen inches and a half long. He is of a fine glossy black throughout, with varying green purple and bronze metallic sheens on the head, neck, and breast.

The female is much smaller, averaging about twelve inches and a half long. The prevailing tone of the bird is snuff brown, darkest on the upper, and shading into obscure buff on the lower parts.

Both sexes have *brown eyes.*

These birds are at all times gregarious, and during the winter are often associated with the Florida Grackles and with the Red-winged Blackbirds in large bands. They are familiar, and when not persecuted very tame and unsuspicious, frequenting wharves and the vicinity of houses near the shore. Their efforts at song are not unpleasing, and on the whole these birds are a very enjoyable and characteristic group in and about the seashore and lake towns of Florida.

Their food is largely insects, small salt and fresh water mollusks, worms, and the like. They also feed on the smaller wild fruits, grain, and seeds. In such pursuits they are very noticeable, strutting along some shore or on some " wild lettuce " bed floating on the water.

In the breeding season they form colonies of varying extent, and in some "saw grass," or bushes on the border of stream or lake, they build their bulky nests. These are made of grasses, seaweed, and like material, wattled into a layer of mud or rotting grass, and lined with fine, dry grasses. From three to five pale bluish or greenish white eggs are laid. These are erratically marked, much in the manner of Bobolink's and Oriole's eggs, with spots, and zigzag lines, and bands of varying shades of brown. They are about an inch and a quarter long and nine tenths of an inch broad. Florida is the metropolis of the Boat-tailed Grackles, but they extend north on the coast as far as Virginia and to Louisiana and Texas.

The Fish Crow is very similar in general appearance to the Common Crow. *It is*, however, *much smaller*, only about sixteen inches long. The **Fish Crow.** birds are black all over, and glossed above with deep blue Corvus ossifragus Wils. or purplish sheen, which is not so obvious below, but is generally much brighter than on this region in the Common Crow. The entire nesting and breeding economy is very much alike in the two birds. The eggs of the bird under consideration are smaller, nearly an inch and a half long by a little more than an inch broad, and are otherwise not appreciably different.

The call notes of the two birds are widely different, that of the Fish Crow recalling the rasping note of the young of the Common Crow, heard so frequently in the early summer.

The Fish Crow is not so widely distributed as its relative. It is not generally found far inland but is distributed over the Atlantic and Gulf Coast region, from Southern Connecticut to Louisiana. It is not uncommon in the lower Hudson Valley, and has been recorded from Massachusetts. In Florida in the winter I have frequently seen large flocks of these birds feeding on the ripe seeds of the cabbage palmetto, and I obtained a single bird in the vicinity of Asheville, North Carolina.

The Kingfisher is the water watchman. In his uniform of blue and white, ever ready to spring his rattle of alarm, he posts himself on some **Belted Kingfisher.** dry limb or other point of vantage from which he may Ceryle alcyon (Linn.). alike watch for his finny prey or the advent of some enemy. These birds are found throughout North America from the Arctic Ocean to Panama. They breed from the Southern States northward and winter from the Middle States south.

The Kingfisher nests in holes in gravel or sand banks, generally excavated by the birds. Such burrows vary from six to eight feet in depth, and the white eggs, four to eight in number, are laid on a rude structure of grasses and trash at the end of the excavation. I have found on a single occasion a pair breeding in a deep crevice in loosened rocks of an old stone quarry. The eggs are about an inch and a third long by a little more than an inch in their other diameter.

BELTED KINGFISHER. ADULT MALE.

The birds are rather more than thirteen inches long, and have very strong beaks (two inches long) and small legs and feet. Both sexes are bluish gray above with some white spotting on the wings. The tail feathers are marked with many spots of white, and indefinitely bordered with the same color. The blue of the upper parts is further broken by a white spot in front of the eye. The feathers of the top of the head are elongated and form a crest when raised. The lower parts are white. There is a band of blue across the breast and the sides are heavily marked with the same color. In the female the breast band is more broken and its color as well as that of the sides is reddish brown. Immature birds are like the adults except for a strong admixture of rusty in the breast band and sides.

The Fish Hawk or Osprey is a bird of almost world wide distribution, having an immense range and inhabiting Europe, Africa, the greater part of Asia, Japan, Formosa, the Australian region, and New Guinea, as well as America. The form which occurs in America has been described as a geographical race of the Old World bird, and is found in Eastern North America, breeding from Florida to Labrador. It winters from South Carolina southward.

American Osprey.
Pandion haliaëtus carolinensis (Gmel.).

At points on the seacoast, as in New Jersey, where they have been carefully protected by *public sentiment*, as well as by legislation, these birds have increased in numbers and have become very familiar and fearless. They breed in the vicinity of houses, even in door yards when the trees afford support for their great nests, and return year after year to the same nesting site. This is in obvious contrast to the account of W. H. Hudson, Esq., at page 215 in his book on British Birds. Speaking of the gradual extinction in Great Britain of this once common bird, he says: "With us it appears in autumn as a migrant in small numbers; but the birds of the British race are now reduced to *one or two pairs* that breed annually in the Highlands of Scotland, and are strictly protected in their summer haunts."

OSPREY OR FISH HAWK.

This harmless and magnificent bird is so fine an adjunct to any sea-shore, is of so great interest to all when at rest near its nest, or when diving from aloft beneath the waves for its finny prey, as to warrant us in using every effort against its destruction. Its care and solicitude in feeding and educating its brood is a matter not soon to be forgotten even by the most casual observer, and its confidence in man is a striking example of how much can be accomplished where *public sentiment* becomes interested, and warrants efforts of a like nature in behalf of other birds.

The Osprey is about two feet long and has an expanse of wing of upwards of five feet. The upper parts are plain grayish brown. The top of the head and back of the neck are much varied with *pure white.* The tail is, like the upper parts grayish brown, is narrowly tipped with white and crossed by six or eight dusky bands. The under parts are white, usually immaculate in the adult male and generally heavily spotted in the female with grayish brown darker than the back. Immature birds have the feathers of the upper parts tipped or bordered with white or buffy, and are otherwise much like the adults.

The nest is an enormous structure of sticks, usually built in a tree from twenty to seventy feet from the ground. The eggs are about two inches and a half long and an inch and four fifths in their smaller diameter. They vary much in appearance from almost white and immaculate to deep buff marked with umber and chocolate, the heavier colors sometimes obscuring the buff or brown.

Adult Duck Hawks are dark slate blue above, becoming appreciably darker on the head. The larger feathers of the wings are barred with deep
Duck Hawk. buff. The tail is obscurely barred with black and each
Falco peregrinus anatum feather has a white tip. The lower parts are white, with a
(Bonap.). cream or buff tinge, barred and marked in a varying degree with dusky brown or black except on the breast.

The sexes are alike in color, but vary much in size, the male being about sixteen inches in length while the female is about three inches longer and proportionately heavier.

Immature birds are dusky brown above, each feather being more or less bordered with buff or rusty brown. There is a blackish area below the eye. The wings are similar to those of the adult bird. The lower parts are deep cream or buff, generally streaked, but sometimes barred with dusky brown or blackish.

The birds nest on shelves on cliffs and in hollows in the higher branches or limbs of trees. The eggs are three or four in number. They vary much in appearance, from buffy marked with reddish brown to deep reddish brown marked with spots of a deeper shade. They are nearly two inches and a tenth long and an inch and three fifths broad.

The Duck Hawk is found in North America and south to Chili in South America. It breeds locally through most of its United States range.

The embodiment of strength in birds, this hawk is the bird of song and story; famed by bard and minstrel, the "Peregrine" of falconry has been celebrated for centuries. Its courage and power command for it a respect that almost causes one to forgive its depredations on poultry and game birds.

Bald Eagle.
Haliæetus leucocephalus
(Linn.).
The male Bald Eagle is about two feet and eight inches long, with a spread of wing of about seven feet. The female is somewhat larger, often exceeding three feet in length and seven feet and a half from tip to tip of the wings. An adult male bird, killed on the Gulf Coast of Florida, in the breeding season, I found to weigh six pounds and three quarters, and an old female taken at the same time and place weighed nine pounds. The Florida birds are not quite as large as more northern representatives.

Adult birds may be readily recognized by the pure white hood extending over the head and neck, and the white tail. The bill and eyes are yellow. The rest of the plumage is dusky brown. Immature birds are entirely dusky brown. This color is generally varied with white or buffy white in infinite variety. The eyes are pale brown and the bill is black. The feathering on the lower leg does not reach to the toes.

The birds nest in trees, generally of considerable height and not far from water. These nests are enormous structures, built of sticks, limbs, and twigs, and are annually repaired; for the birds, if not disturbed or destroyed, use the same nest for many years. The eggs, two or three in number, are white and immaculate. They are about two inches and nine tenths long by a little more than two inches and a quarter broad.

The Bald Eagle is found throughout North America, generally in the vicinity of water, and breeds throughout its range.

These eagles live largely on fish, often such as are cast up on the beach and frequently those stolen from the Osprey. They are also fond of carrion, and I have seen them in the South feeding on the dead carcasses of cattle or alligators, in company with the two kinds of Buzzards and the Caracara Eagle.

BALD EAGLE.

The Gray Sea Eagle is a common bird in Southeastern Greenland, and an inhabitant of the northern part of the Eastern Hemisphere.

Gray Sea Eagle.
Haliæetus albicilla (Linn.).
The adult bird is a little larger and in general appearance is not unlike our Bald Eagle, but has the head brown, like the rest of the plumage, but somewhat lighter in tone. The *tail is pure white*. The bill and eyes are yellow.

The birds nest on cliffs or tall trees generally near water, much as do our Bald Eagles, and the eggs, two or three in number, are similar in color, but a little larger.

The Snail Hawk or Everglade Kite reminds one, at first glance, of a Marsh Harrier. The male birds are nearly as large, being about eighteen

Everglade Kite.
Rostrhamus sociabilis
(Vieill.).
inches long ; the females are some two inches longer. They are slimly built and not heavy bodied, with a comparatively large spread of wing and long tails and very hooked and slender beaks and claws.

The one region where I have met with these birds in numbers is Panasoffkee Lake in Sumpter County, Florida. Here, in the early spring of 1876, they arrived from the south in February, and soon became very abundant and gregarious, from two to twenty being associated together. Their food consisted of a kind of large fresh water snail, which is very abundant in this lake. These snails were caught by the birds, who fished for them over shallow water, reminding one of gulls. Their prey was secured by diving, and immediately taken to some frequented perch near by. With dexterity the birds removed the snails from their shells without breaking the shell or the operculum that closed it. Great piles of these shells below such perches attested to the enormous number of snails consumed.

Probably not less than two hundred pairs bred here, but up to the time of my departure on March 25th, they had not laid, though nest building had begun.

The male is a very dark slate color, relieved by a white patch above and a smaller white patch below the tail. The tail is composed of feathers, white at their bases, shading into darker slate than the body, and tipped rather narrowly with white. The eyes are red. The female and immature birds are dusky brown above, each feather being tipped with chestnut. The lower parts are barred, streaked, and spotted with brown, reddish brown, buff, and dusky, giving, on the whole, a mottled appearance. The wings are dusky, and the tail is much as in the adult male, the same white patches above and below being noticeable. The eyes in the female and immature birds are brown.

The birds nest in bushes near to the ground and generally in "saw grass" marshes. The eggs are two or three in number, with a ground color of pale bluish white, heavily spotted and often washed with reddish and chocolate brown. They are about an inch and three quarters long and somewhat less than an inch and a half broad.

The birds, so far as ascertained, occur only in Florida in the United States, where they breed locally in the southern half of the State, migrating south in the winter. They are common in localities in Cuba, Eastern Mexico, and Central America. They also occur in Eastern South America, south to the Argentine Republic.

A SYLVAN STREAM.

A SYSTEMATIC TABLE OF THE LAND BIRDS OF EASTERN NORTH AMERICA.

ORDER GALLINÆ. GALLINACEOUS BIRDS.

SUBORDER PHASIANI. PHEASANTS, GROUSE, PARTRIDGES, QUAILS, ETC.

FAMILY TETRAONIDÆ. GROUSE, PARTRIDGES, ETC.

SUBFAMILY PERDICINÆ. PARTRIDGES.

GENUS COTURNIX BONNATERRE.

Coturnix coturnix (Linn.).
Coturnix coturnix (Linn.).
European Quail.

GENUS COLINUS LESSON.

Colinus virginianus (Linn.).
Bob-white.

Colinus virginianus floridanus (Coues).
Florida Bob-white.

SUBFAMILY TETRAONINÆ. GROUSE.

GENUS DENDRAGAPUS ELLIOT.

Dendragapus canadensis (Linn.).
Canada Grouse.

GENUS TETRAO LINNÆUS.

Tetrao tetrix Linn.
Black Cock.

GENUS BONASA STEPHENS.

Bonasa umbellus (Linn.).
Ruffed Grouse.

Bonasa umbellus togata (Linn.).
Canadian Ruffed Grouse.

GENUS LAGOPUS BRISSON.

Lagopus lagopus (Linn.).
Willow Ptarmigan.

Lagopus lagopus alleni Stejn.
Allen's Ptarmigan.

Lagopus rupestris (Gmel.).
Rock Ptarmigan.

Lagopus rupestris reinhardi (Brehm).
Reinhardt's Ptarmigan.

Lagopus welchi Brewst.
Welch's Ptarmigan.

329

ORDER GALLINÆ.   GALLINACEOUS BIRDS.   (*Continued.*)

SUBORDER PHASIANI.   PHEASANTS, GROUSE, PARTRIDGES, QUAILS, ETC.   (*Continued.*)

FAMILY **TETRAONIDÆ.** GROUSE, PARTRIDGES, ETC. (*Continued.*)

SUBFAMILY **TETRAONINÆ.** (*Continued.*)

GENUS TYMPANUCHUS GLOGER.

Tympanuchus americanus (Reich.).
  Prairie Hen.

Tympanuchus cupido (Linn.).
  Heath Hen.

GENUS PEDIOCÆTES BAIRD.

Pediocætes phasianellus campestris Ridgw.
  Prairie Sharp-tailed Grouse.

FAMILY **PHASIANIDÆ.** PHEASANTS, ETC.

SUBFAMILY **PHASIANINÆ.** PHEASANTS.

GENUS PHASIANUS LINNÆUS.

Phasianus colchicus Linn.
  Pheasant.

SUBFAMILY **MELEAGRINÆ.** TURKEYS

GENUS MELEAGRIS LINNÆUS.

Meleagris gallopavo Linn.
  Wild Turkey.

Meleagris gallopavo osceola Scott.
  Florida Wild Turkey.

ORDER COLUMBÆ.   PIGEONS.

FAMILY **COLUMBIDÆ.** PIGEONS.

GENUS COLUMBA LINNÆUS.

Columba leucocephala Linn.
  White-crowned Pigeon.

GENUS ECTOPISTES SWAINSON.

Ectopistes migratorius (Linn.).
  Passenger Pigeon.

GENUS ZENAIDURA BONAPARTE.

Zenaidura macroura (Linn.).
  Mourning Dove.

GENUS ZENAIDA BONAPARTE.

Zenaida zenaida (Bonap.).
  Zenaida Dove.

GENUS MELOPELIA BONAPARTE.

Melopelia leucoptera (Linn.).
  White-winged Dove.

GENUS COLUMBIGALLINA BOIE.

Columbigallina passerina terrestris Chapm.
  Ground Dove.

Land Birds of Eastern North America.

ORDER COLUMBÆ.   PIGEONS.   (*Continued.*)

FAMILY **COLUMBIDÆ.**   PIGEONS.   (*Continued.*)

GENUS GEOTRYGON GOSSE.

Geotrygon chrysia Bonap.
  Key West Quail-Dove.

Geotrygon montana (Linn.).
  Ruddy Quail-Dove.

GENUS STARNŒNAS BONAPARTE.

Starnœnas cyanocephala (Linn.).
  Blue-Headed Quail-Dove.

ORDER RAPTORES.   BIRDS OF PREY.

SUBORDER SARCORHAMPHI.   AMERICAN VULTURES.

FAMILY **CATHARTIDÆ.**   AMERICAN VULTURES.

GENUS CATHARTES ILLIGER.

Cathartes aura (Linn.).
  Turkey Vulture.

GENUS CATHARISTA VIEILLOT.

Catharista atrata (Bartr.).
  Black Vulture.

SUBORDER FALCONES.   VULTURES, FALCONS, HAWKS, BUZZARDS, EAGLES, KITES, HARRIERS, ETC.

FAMILY **FALCONIDÆ.**   VULTURES, FALCONS, HAWKS, EAGLES.

SUBFAMILY **ACCIPITRINÆ.**   KITES, BUZZARDS, HAWKS, GOSHAWKS, EAGLES.

GENUS ELANOIDES VIEILLOT.

Elanoides forficatus (Linn.).
  Swallow-tailed Kite.

GENUS ELANUS SAVIGNY.

Elanus leucurus (Vieill.).
  White-tailed Kite.

GENUS ICTINIA VIEILLOT.

Ictinia mississippiensis (Wils.).
  Mississippi Kite.

GENUS ROSTRHAMUS LESSON.

Rostrhamus sociabilis (Vieill.).
  Everglade Kite.

GENUS CIRCUS LACÉPÈDE.

Circus hudsonius (Linn.).
  Marsh Hawk.

GENUS ACCIPITER BRISSON.

Accipiter velox (Wils.).
  Sharp-shinned Hawk.

Bird Studies.

ORDER RAPTORES. BIRDS OF PREY. (*Continued.*)

SUBORDER FALCONES. VULTURES, FALCONS, HAWKS, BUZZARDS, EAGLES, KITES, HARRIERS, ETC. (*Continued.*)

FAMILY **FALCONIDÆ.** VULTURES, FALCONS, HAWKS, EAGLES. (*Continued.*)

SUBFAMILY **ACCIPITRINÆ.** KITES, BUZZARDS, HAWKS, GOSHAWKS, EAGLES. (*Continued.*)

GENUS ACCIPITER BRISSON. (*Continued.*)

Accipiter cooperii (Bonap.).
  Cooper's Hawk.

Accipiter atricapillus (Wils.).
  American Goshawk.

GENUS BUTEO CUVIER.

Buteo buteo (Linn.).
  European Buzzard.

Buteo borealis (Gmel.).
  Red-tailed Hawk.

Buteo borealis kriderii Hoopes.
  Krider's Hawk.

Buteo borealis calurus (Cass.).
  Western Red-tail.

Buteo borealis harlani (Aud.).
  Harlan's Hawk.

Buteo lineatus (Gmel.).
  Red-shouldered Hawk.

Buteo lineatus alleni Ridgw.
  Florida Red-shouldered Hawk.

Buteo swainsoni Bonap.
  Swainson's Hawk.

Buteo latissimus (Wils.).
  Broad-winged Hawk.

Buteo brachyurus Vieill.
  Short-tailed Hawk.

GENUS ASTURINA VIEILLOT.

Asturina plagiata Schlegel.
  Mexican Goshawk.

GENUS ARCHIBUTEO BREHM.

Archibuteo lagopus sancti-johannis (Gmel.).
  American Rough-legged Hawk.

Archibuteo ferrugineus (Licht.).
  Ferruginous Rough-Leg.

GENUS AQUILA BRISSON.

Aquila chrysaëtos (Linn.).
  Golden Eagle.

GENUS HALIÆETUS SAVIGNY.

Haliæetus albicilla (Linn.).
  Gray Sea Eagle.

Land Birds of Eastern North America.

ORDER RAPTORES. BIRDS OF PREY. (*Continued.*)

SUBORDER FALCONES. VULTURES, FALCONS, HAWKS, BUZZARDS, EAGLES, KITES, HARRIERS, ETC. (*Continued.*)

FAMILY **FALCONIDÆ.** VULTURES, FALCONS, HAWKS, EAGLES. (*Continued.*)

SUBFAMILY **ACCIPITRINÆ.** KITES, BUZZARDS, HAWKS, GOSHAWKS, EAGLES. (*Continued.*)

GENUS HALIÆETUS SAVIGNY. (*Continued.*)

Haliæetus leucocephalus (Linn.).
  Bald Eagle.

SUBFAMILY **FALCONINÆ.** FALCONS.

GENUS FALCO LINNÆUS.

Falco islandus Brünn.
  White Gyrfalcon.

Falco rusticolus Linn.
  Gray Gyrfalcon.

Falco rusticolus gyrfalco (Linn.).
  Gyrfalcon.

Falco rusticolus obsoletus (Gmel.).
  Black Gyrfalcon.

Falco mexicanus Schleg.
  Prairie Falcon.

Falco peregrinus anatum (Bonap.).
  Duck Hawk.

Falco columbarius Linn.
  Pigeon Hawk.

Falco regulus Pall.
  Merlin.

Falco tinnunculus Linn.
  Kestrel.

Falco sparverius Linn.
  American Sparrow Hawk.

Falco dominicensis Gmel.
  Cuban Sparrow Hawk.

GENUS POLYBORUS VIEILLOT.

Polyborus cheriway (Jacq.).
  Audubon's Caracara.

SUBFAMILY **PANDIONINÆ.** OSPREYS.

GENUS PANDION SAVIGNY.

Pandion haliaëtus carolinensis (Gmel.).
  American Osprey.

SUBORDER STRIGES. OWLS.

FAMILY **STRIGIDÆ.** BARN OWLS.

GENUS STRIX LINNÆUS.

Strix pratincola Bonap.
  American Barn Owl.

ORDER RAPTORES. BIRDS OF PREY. (*Continued.*)

SUBORDER STRIGES. OWLS. (*Continued.*)

FAMILY **BUBONIDÆ.** HORNED OWLS, ETC.

GENUS ASIO BRISSON.
Asio wilsonianus (Less.).
American Long-eared Owl.
Asio accipitrinus (Pall.).
Short-eared Owl.

GENUS SYRNIUM SAVIGNY.
Syrnium nebulosum (Forst.).
Barred Owl.
Syrnium nebulosum alleni Ridgw.
Florida Barred Owl.

GENUS SCOTIAPTEX SWAINSON.
Scotiaptex cinerea (Gmel.).
Great Gray Owl.

GENUS NYCTALA BREHM.
Nyctala tengmalmi richardsoni (Bonap.).
Richardson's Owl.
Nyctala acadica (Gmel.).
Saw-whet Owl.

GENUS MEGASCOPS KAUP.
Megascops asio (Linn.).
Screech Owl.
Megascops asio floridanus (Ridgw.).
Florida Screech Owl.

GENUS BUBO DUMÉRIL.
Bubo virginianus (Gmel.).
Great Horned Owl.
Bubo virginianus subarcticus (Hoy).
Western Horned Owl.
Bubo virginianus saturatus Ridgw.
Dusky Horned Owl.

GENUS NYCTEA STEPHENS.
Nyctea nyctea (Linn.).
Snowy Owl.

GENUS SURNIA DUMÉRIL.
Surnia ulula caparoch (Müll.).
American Hawk Owl.

GENUS SPEOTYTO GLOGER.
Speotyto cunicularia hypogæa (Bonap.).
Burrowing Owl.
Speotyto cunicularia floridana Ridgw.
Florida Burrowing Owl.

ORDER PSITTACI. PARROTS, MACAWS, PAROQUETS, ETC.

FAMILY **PSITTACIDÆ.** PARROTS AND PAROQUETS.
GENUS CONURUS KUHL.
Conurus carolinensis (Linn.).
Carolina Paroquet.

ORDER COCCYGES. CUCKOOS, ETC.

SUBORDER CUCULI. CUCKOOS, ETC.

FAMILY **CUCULIDÆ.** CUCKOOS, ANIS, ETC.
SUBFAMILY **CROTOPHAGINÆ.** ANIS.
GENUS CROTOPHAGA LINNÆUS.
Crotophaga ani Linn.
Ani.
SUBFAMILY **COCCYGINÆ.** AMERICAN CUCKOOS.
GENUS COCCYZUS VIEILLOT.
Coccyzus minor (Gmel.).
Mangrove Cuckoo.
Coccyzus minor maynardi (Ridgw.).
Maynard's Cuckoo.
Coccyzus americanus (Linn.).
Yellow-billed Cuckoo.
Coccyzus erythrophthalmus (Wils.).
Black-billed Cuckoo.

SUBORDER ALCYONES. KINGFISHERS.

FAMILY **ALCEDINIDÆ.** KINGFISHERS.
GENUS CERYLE BOIE.
Ceryle alcyon (Linn.).
Belted Kingfisher.

ORDER PICI. WOODPECKERS, WRYNECKS, ETC.

FAMILY **PICIDÆ.** WOODPECKERS.
GENUS CAMPEPHILUS GRAY.
Campephilus principalis (Linn.).
Ivory-billed Woodpecker.
GENUS DRYOBATES BOIE.
Dryobates villosus (Linn.).
Hairy Woodpecker.
Dryobates villosus leucomelas (Bodd.).
Northern Hairy Woodpecker.

ORDER PICI. WOODPECKERS, WRYNECKS, ETC. (*Continued.*)

FAMILY **PICIDÆ.** WOODPECKERS. (*Continued.*)

GENUS DRYOBATES BOIE. (*Continued.*)

Dryobates villosus audubonii (Swains.).
Southern Hairy Woodpecker.

Dryobates pubescens (Linn.).
Southern Downy Woodpecker.

Dryobates pubescens medianus (Swains.).
Downy Woodpecker.

Dryobates borealis (Vieill.).
Red-cockaded Woodpecker.

GENUS PICOIDES LACÉPÈDE.

Picoides arcticus (Swains.).
Arctic Three-toed Woodpecker.

Picoides americanus Brehm.
American Three-toed Woodpecker.

GENUS SPHYRAPICUS BAIRD.

Sphyrapicus varius (Linn.).
Yellow-bellied Woodpecker.

GENUS CEOPHLŒUS CABANIS.

Ceophlœus pileatus (Linn.).
Pileated Woodpecker.

GENUS MELANERPES SWAINSON.

Melanerpes erythrocephalus (Linn.).
Red-headed Woodpecker.

Melanerpes carolinus (Linn.).
Red-bellied Woodpecker.

GENUS COLAPTES SWAINSON.

Colaptes auratus (Linn.).
Flicker.

ORDER MACROCHIRES. GOATSUCKERS, SWIFTS, ETC.

SUBORDER CAPRIMULGI. GOATSUCKERS, ETC.

FAMILY **CAPRIMULGIDÆ.** GOATSUCKERS, ETC.

GENUS ANTROSTOMUS GOULD.

Antrostomus carolinensis (Gmel.).
Chuck-will's-widow.

Antrostomus vociferus (Wils.).
Whip-poor-will.

GENUS CHORDEILES SWAINSON.

Chordeiles virginianus (Gmel.).
Nighthawk.

Chordeiles virginianus henryi (Cass.).
Western Nighthawk.

Chordeiles virginianus chapmani (Coues).
Florida Nighthawk.

ORDER MACROCHIRES. GOATSUCKERS, SWIFTS, ETC. (*Continued.*)

SUBORDER CYPSELI. SWIFTS.

FAMILY **MICROPODIDÆ.** SWIFTS.

SUBFAMILY **CHÆTURINÆ.** SPINE-TAILED SWIFTS.

GENUS CHÆTURA STEPHENS.

Chætura pelagica (Linn.).
Chimney Swift.

SUBORDER TROCHILI. HUMMINGBIRDS.

FAMILY **TROCHILIDÆ.** HUMMINGBIRDS.

GENUS TROCHILUS LINNÆUS.

Trochilus colubris Linn.
Ruby-throated Hummingbird.

ORDER PASSERES. PERCHING BIRDS.

SUBORDER CLAMATORES. SONGLESS PERCHING BIRDS.

FAMILY **TYRANNIDÆ.** TYRANT FLYCATCHERS.

GENUS MILVULUS SWAINSON.

Milvulus tyrannus (Linn.).
Fork-tailed Flycatcher.

Milvulus forficatus (Gmel.).
Scissor-tailed Flycatcher.

GENUS TYRANNUS CUVIER.

Tyrannus tyrannus (Linn.).
Kingbird.

Tyrannus dominicensis (Gmel.).
Gray Kingbird.

Tyrannus verticalis Say.
Arkansas Kingbird.

GENUS MYIARCHUS CABANIS.

Myiarchus crinitus (Linn.).
Crested Flycatcher.

GENUS SAYORNIS BONAPARTE.

Sayornis phœbe (Lath.).
Phœbe.

Sayornis saya (Bonap.).
Say's Phœbe.

GENUS CONTOPUS CABANIS.

Contopus borealis (Swains.).
Olive-sided Flycatcher.

Contopus virens (Linn.).
Wood Pewee.

ORDER PASSERES. PERCHING BIRDS. (*Continued.*)

SUBORDER CLAMATORES. SONGLESS PERCHING BIRDS. (*Continued.*)

FAMILY **TYRANNIDÆ.** TYRANT FLYCATCHERS. (*Continued.*)

GENUS EMPIDONAX CABANIS.

Empidonax flaviventris Baird.
Yellow-bellied Flycatcher.

Empidonax virescens (Vieill.).
Green-crested Flycatcher.

Empidonax traillii (Aud.).
Traill's Flycatcher.

Empidonax traillii alnorum Brewst.
Alder Flycatcher.

Empidonax minimus Baird.
Least Flycatcher.

SUBORDER OSCINES. SONG BIRDS

FAMILY **ALAUDIDÆ.** LARKS.

GENUS ALAUDA LINNÆUS.

Alauda arvensis Linn.
Skylark.

GENUS OTOCORIS BONAPARTE.

Otocoris alpestris (Linn.).
Horned Lark.

Otocoris alpestris praticola Hensh.
Prairie Horned Lark.

FAMILY **CORVIDÆ.** CROWS, JAYS, MAGPIES, ETC.

SURFAMILY **GARRULINÆ.** MAGPIES AND JAYS.

GENUS PICA BRISSON.

Pica pica hudsonica (Sab.).
American Magpie.

GENUS CYANOCITTA STRICKLAND.

Cyanocitta cristata (Linn.).
Blue Jay.

Cyanocitta cristata florincola Coues.
Florida Blue Jay.

GENUS APHELOCOMA CABANIS.

Aphelocoma floridana (Bartr.).
Florida Jay.

GENUS PERISOREUS BONAPARTE.

Perisoreus canadensis (Linn.).
Canada Jay.

Perisoreus canadensis nigricapillus Ridgw.
Labrador Jay.

Land Birds of Eastern North America.

ORDER PASSERES. PERCHING BIRDS. (*Continued.*)

SUBORDER OSCINES. SONG BIRDS. (*Continued.*)

FAMILY **CORVIDÆ.** CROWS, JAYS, MAGPIES, ETC. (*Continued.*)

SUBFAMILY **CORVINÆ.** CROWS.

GENUS CORVUS LINNÆUS.

Corvus corax principalis Ridgw.
Northern Raven.

Corvus americanus Aud.
American Crow.

Corvus americanus floridanus Baird.
Florida Crow.

Corvus ossifragus Wils.
Fish Crow.

FAMILY **STURNIDÆ.** STARLINGS.

GENUS STURNUS LINNÆUS.

Sturnus vulgaris Linn.
Starling.

FAMILY **ICTERIDÆ.** BLACKBIRDS, ORIOLES, ETC.

GENUS DOLICHONYX SWAINSON.

Dolichonyx oryzivorus (Linn.).
Bobolink.

GENUS MOLOTHRUS SWAINSON.

Molothrus ater (Bodd.).
Cowbird.

GENUS XANTHOCEPHALUS BONAPARTE.

Xanthocephalus xanthocephalus (Bonap.).
Yellow-headed Blackbird.

GENUS AGELAIUS Vieillot.

Agelaius phœniceus (Linn.).
Red-winged Blackbird.

Agelaius phœniceus floridanus Maynard.
Florida Red-wing.

GENUS STURNELLA Vieillot.

Sturnella magna (Linn.).
Meadowlark.

Sturnella magna neglecta (Aud.).
Western Meadowlark.

GENUS ICTERUS BRISSON.

Icterus icterus (Linn.).
Troupial.

Icterus spurius (Linn.).
Orchard Oriole.

ORDER PASSERES. PERCHING BIRDS. (*Continued.*)

SUBORDER OSCINES. SONG BIRDS. (*Continued.*)

FAMILY **ICTERIDÆ.** BLACKBIRDS, ORIOLES, ETC. (*Continued.*)
GENUS ICTERUS BRISSON. (*Continued.*)
Icterus galbula (Linn.).
　Baltimore Oriole.
· Icterus bullocki (Swains.).
　Bullock's Oriole.
GENUS SCOLECOPHAGUS SWAINSON.
Scolecophagus carolinus (Müll.).
　Rusty Blackbird.
Scolecophagus cyanocephalus (Wagl.).
　Brewer's Blackbird.
GENUS QUISCALUS VIEILLOT.
Quiscalus quiscula (Linn.).
　Purple Grackle.
Quiscalus quiscula aglæus (Baird).
　Florida Grackle.
Quiscalus quiscula æneus (Ridgw.).
　Bronzed Grackle.
Quiscalus major Vieill.
　Boat-tailed Grackle.

FAMILY **FRINGILLIDÆ.** FINCHES, SPARROWS, ETC.
GENUS COCCOTHRAUSTES BRISSON.
Coccothraustes vespertinus (Coop.).
　Evening Grosbeak.

GENUS PINICOLA VIEILLOT.
Pinicola enucleator (Linn.).
　Pine Grosbeak.

GENUS CARPODACUS KAUP.
Carpodacus purpureus (Gmel.).
　Purple Finch.

GENUS LOXIA LINNÆUS.
Loxia curvirostra minor (Brehm).
　American Crossbill.
Loxia leucoptera Gmel.
　White-winged Crossbill.

GENUS ACANTHIS BECHSTEIN.
Acanthis hornemannii (Holb.).
　Greenland Redpoll.
Acanthis hornemannii exilipes (Coues).
　Hoary Redpoll.
Acanthis linaria (Linn.).
　Redpoll.
Acanthis linaria holbœllii (Brehm).
　Holböll's Redpoll.

Land Birds of Eastern North America. 341

ORDER PASSERES. PERCHING BIRDS. (*Continued.*)

SUBORDER OSCINES. SONG BIRDS. (*Continued.*)

FAMILY **FRINGILLIDÆ.** FINCHES, SPARROWS, ETC. (*Continued.*)
GENUS ACANTHIS BECHSTEIN. (*Continued.*)
Acanthis linaria rostrata (Coues).
　Greater Redpoll.

GENUS SPINUS KOCH.
Spinus tristis (Linn.).
　American Goldfinch.

Spinus pinus (Wils.).
　Pine Siskin.

GENUS CARDUELIS BRISSON.
Carduelis carduelis (Linn.).
　European Goldfinch.

GENUS PASSER BRISSON.
Passer domesticus (Linn.).
　House Sparrow.

Passer montanus (Linn.).
　European Tree Sparrow.

GENUS PLECTROPHENAX STEJNEGER.
Plectrophenax nivalis (Linn.).
　Snowflake.

GENUS CALCARIUS BECHSTEIN.
Calcarius lapponicus (Linn.).
　Lapland Longspur.

Calcarius pictus (Swains.).
　Smith's Longspur.

Calcarius ornatus (Towns.).
　Chestnut-collared Longspur.

GENUS RHYNCHOPHANES BAIRD.
Rhynchophanes mccownii (Lawr.).
　McCown's Longspur.

GENUS POOCÆTES BAIRD.
Poocætes gramineus (Gmel.).
　Vesper Sparrow.

GENUS AMMODRAMUS SWAINSON.
Ammodramus princeps (Mayn.).
　Ipswich Sparrow.

Ammodramus sandwichensis savanna (Wils.).
　Savanna Sparrow.

Ammodramus savannarum passerinus (Wils.).
　Grasshopper Sparrow.

Ammodramus henslowii (Aud.).
　Henslow's Sparrow.

Bird Studies.

ORDER PASSERES. PERCHING BIRDS. (*Continued.*)

SUBORDER OSCINES. SONG BIRDS. (*Continued.*)

FAMILY **FRINGILLIDÆ.** FINCHES, SPARROWS, ETC. (*Continued.*)
GENUS AMMODRAMUS SWAINSON. (*Continued.*)

Ammodramus leconteii (Aud.).
 Leconte's Sparrow.

Ammodramus caudacutus (Gmel.).
 Sharp-tailed Sparrow.

Ammodramus caudacutus nelsoni Allen.
 Nelson's Sparrow.

Ammodramus caudacutus subvirgatus Dwight.
 Acadian Sharp-tailed Sparrow.

Ammodramus maritimus (Wils.).
 Seaside Sparrow.

Ammodramus maritimus peninsulæ Allen.
 Scott's Seaside Sparrow.

Ammodramus maritimus macgillivrayi Aud.
 Louisiana Seaside Sparrow.

Ammodramus nigrescens Ridgw.
 Dusky Seaside Sparrow.

GENUS CHONDESTES SWAINSON.

Chondestes grammacus (Say).
 Lark Sparrow.

GENUS ZONOTRICHIA SWAINSON.

Zonotrichia leucophrys (Forst.).
 White-crowned Sparrow.

Zonotrichia albicollis (Gmel.).
 White-throated Sparrow.

GENUS SPIZELLA BONAPARTE.

Spizella monticola (Gmel.).
 Tree Sparrow.

Spizella socialis (Wils.).
 Chipping Sparrow.

Spizella pallida (Swains.).
 Clay-colored Sparrow.

Spizella breweri Cass.
 Brewer's Sparrow.

Spizella pusilla (Wils.).
 Field Sparrow.

Spizella pusilla arenacea Chadb.
 Western Field Sparrow.

GENUS JUNCO WAGLER.

Junco hyemalis (Linn.).
 Slate-colored Snow-bird.

Junco hyemalis connectens Coues.
 Hybrid Snow-bird.

ORDER PASSERES. PERCHING BIRDS. (*Continued.*)

SUBORDER OSCINES. SONG BIRDS. (*Continued.*)

FAMILY **FRINGILLIDÆ.** FINCHES, SPARROWS, ETC. (*Continued.*)

GENUS JUNCO WAGLER. (*Continued.*)

Junco hyemalis carolinensis Brewst.
Carolina Snow-bird.

GENUS PEUCÆA AUDUBON.

Peucæa æstivalis (Licht.).
Pine-woods Sparrow.

Peucæa æstivalis bachmanii (Aud.).
Bachman's Sparrow.

GENUS MELOSPIZA BAIRD.

Melospiza fasciata (Gmel.).
Song Sparrow.

Melospiza lincolnii (Aud.).
Lincoln's Sparrow.

Melospiza georgiana (Lath.).
Swamp Sparrow.

GENUS PASSERELLA SWAINSON.

Passerella iliaca (Merr.).
Fox Sparrow.

GENUS PIPILO VIEILLOT.

Pipilo erythrophthalmus (Linn.).
Towhee.

Pipilo erythrophthalmus alleni Coues.
White-eyed Towhee.

GENUS CARDINALIS BONAPARTE.

Cardinalis cardinalis (Linn.).
Cardinal.

Cardinalis cardinalis floridanus Ridgw.
Florida Cardinal.

GENUS ZAMELODIA COUES.

Zamelodia ludoviciana (Linn.).
Rose-breasted Grosbeak.

Zamelodia melanocephala (Swains.).
Black-headed Grosbeak.

GENUS GUIRACA SWAINSON.

Guiraca cærulea (Linn.).
Blue Grosbeak.

GENUS PASSERINA VIEILLOT.

Passerina cyanea (Linn.).
Indigo Bunting.

Passerina versicolor (Bonap.).
Varied Bunting.

Passerina ciris (Linn.).
Nonpareil.

ORDER PASSERES. PERCHING BIRDS. (*Continued*)

SUBORDER OSCINES. SONG BIRDS. (*Continued.*)

FAMILY **FRINGILLIDÆ.** FINCHES, SPARROWS, ETC. (*Continued.*)
GENUS EUETHEIA REICHENBACH.
Euetheia bicolor (Linn.).
     Grassquit.
Euetheia canora (Gmel.).
     Melodious Grassquit.
GENUS SPIZA BONAPARTE.
Spiza americana (Gmel.).
     Dickcissel.
GENUS CALAMOSPIZA BONAPARTE.
Calamospiza melanocorys Stejn.
     Lark Bunting.

FAMILY **TANAGRIDÆ.** TANAGERS.
GENUS PIRANGA VIEILLOT.
Piranga ludoviciana (Wils.).
     Louisiana Tanager.
Piranga erythromelas Vieill.
     Scarlet Tanager.
Piranga rubra (Linn.).
     Summer Tanager.

FAMILY **HIRUNDINIDÆ.** SWALLOWS.
GENUS PROGNE BOIE.
Progne subis (Linn.).
     Purple Martin.
Progne cryptoleuca Baird.
     Cuban Martin.
GENUS PETROCHELIDON CABANIS.
Petrochelidon lunifrons (Say).
     Cliff Swallow.
Petrochelidon fulva (Vieill.).
     Cuban Cliff Swallow.

GENUS CHELIDON FORSTER.
Chelidon erythrogastra (Bodd.).
     Barn Swallow.
GENUS TACHYCINETA CABANIS.
Tachycineta bicolor (Vieill.).
     Tree Swallow.

GENUS CALLICHELIDON BAIRD.
Callichelidon cyaneoviridis (Bryant).
     Bahaman Swallow.
GENUS CLIVICOLA FORSTER.
Clivicola riparia (Linn.).
     Bank Swallow.

ORDER PASSERES. PERCHING BIRDS. (*Continued.*)

SUBORDER OSCINES. SONG BIRDS. (*Continued.*)

FAMILY **HIRUNDINIDÆ.** SWALLOWS. (*Continued.*)
GENUS STELGIDOPTERYX BAIRD.
Stelgidopteryx serripennis (Aud.).
Rough-winged Swallow.

FAMILY **AMPELIDÆ.** WAXWINGS, ETC.

SUBFAMILY AMPELINÆ. WAXWINGS.
GENUS AMPELIS LINNÆUS.
Ampelis garrulus Linn.
Bohemian Waxwing.
Ampelis cedrorum (Vieill.).
Cedar Waxwing.

FAMILY **LANIIDÆ.** SHRIKES.
GENUS LANIUS LINNÆUS.

Lanius borealis Vieill.
Northern Shrike.

Lanius ludovicianus Linn.
Loggerhead Shrike.

FAMILY **VIREONIDÆ.** VIREOS.
GENUS VIREO VIEILLOT.

Vireo calidris barbatulus (Cab.).
Black-whiskered Vireo.
Vireo olivaceus (Linn.).
Red-eyed Vireo.
Vireo flavoviridis (Cass.).
Yellow-green Vireo.
Vireo philadelphicus (Cass.).
Philadelphia Vireo.
Vireo gilvus (Vieill.).
Warbling Vireo.
Vireo flavifrons Vieill.
Yellow-throated Vireo.
Vireo solitarius (Wils.).
Blue-headed Vireo.
Vireo solitarius plumbeus (Coues).
Plumbeous Vireo.
Vireo solitarius alticola Brewst.
Mountain Solitary Vireo.
Vireo noveboracensis (Gmel.).
White-eyed Vireo.
Vireo noveboracensis maynardi Brewst.
Key West Vireo.

346 Bird Studies.

FAMILY **VIREONIDÆ.** VIREOS. (*Continued.*)
GENUS VIREO VIEILLOT. (*Continued.*)
Vireo bellii Aud.
    Bell's Vireo.
FAMILY **CŒREBIDÆ.** HONEY CREEPERS.
GENUS CŒREBA VIEILLOT.
Cœreba bahamensis (Reich.).
    Bahama Honey Creeper.
FAMILY **MNIOTILTIDÆ.** WOOD WARBLERS.
GENUS MNIOTILTA VIEILLOT.
Mniotilta varia (Linn.).
    Black and White Warbler.
GENUS PROTONOTARIA BAIRD.
Protonotaria citrea (Bodd.).
    Prothonotary Warbler.
GENUS HELINAIA AUDUBON.
Helinaia swainsonii Aud.
    Swainson's Warbler.
GENUS HELMITHERUS RAFINESQUE.
Helmitherus vermivorus (Gmel.).
    Worm-eating Warbler.
GENUS HELMINTHOPHILA RIDGWAY.
Helminthophila bachmanii (Aud.).
    Bachman's Warbler.
Helminthophila pinus (Linn.).
    Blue-winged Warbler.
Helminthophila chrysoptera (Linn.).
    Golden-winged Warbler.
Helminthophila lawrencei (Herrick).
    Lawrence's Warbler.
Helminthophila leucobronchialis (Brewst.).
    Brewster's Warbler.
Helminthophila rubricapilla (Wils.).
    Nashville Warbler.
Helminthophila celata (Say).
    Orange-crowned Warbler.
Helminthophila peregrina (Wils.).
    Tennessee Warbler.
GENUS COMPSOTHLYPIS CABANIS.
Compsothlypis americana (Linn.).
    Parula Warbler.
Compsothlypis americana usneæ Brewster.
    Northern Parula Warbler.

ORDER PASSERES. PERCHING BIRDS. (*Continued.*)

SUBORDER OSCINES. SONG BIRDS. (*Continued.*)

FAMILY **MNIOTILTIDÆ.** WOOD WARBLERS. (*Continued.*)

GENUS DENDROICA GRAY.

Dendroica tigrina (Gmel.).
Cape May Warbler.

Dendroica æstiva (Gmel.).
Yellow Warbler.

Dendroica cærulescens (Gmel.).
Black-throated Blue Warbler.

Dendroica cærulescens cairnsi Coues.
Cairns's Warbler.

Dendroica coronata (Linn.).
Myrtle Warbler.

Dendroica auduboni (Towns.).
Audubon's Warbler.

Dendroica maculosa (GMEL.).
Magnolia Warbler.

Dendroica rara (Wils.).
Cerulean Warbler.

Dendroica pensylvanica (Linn.).
Chestnut-sided Warbler.

Dendroica castanea (Wils.).
Bay-breasted Warbler.

Dendroica striata (Forst.).
Black-poll Warbler.

Dendroica blackburniæ (Gmel.).
Blackburnian Warbler.

Dendroica dominica (Linn.).
Yellow-throated Warbler.

Dendroica dominica albilora Ridgw.
Sycamore Warbler.

Dendroica virens (Gmel.).
Black-throated Green Warbler.

Dendroica townsendi (Towns.).
Townsend's Warbler.

Dendroica kirtlandi Baird.
Kirtland's Warbler.

Dendroica vigorsii (Aud.).
Pine Warbler.

Dendroica palmarum (Gmel.).
Palm Warbler.

Dendroica palmarum hypochrysea Ridgw.
Yellow Palm Warbler.

Dendroica discolor (Vieill.).
Prairie Warbler.

GENUS SEIURUS SWAINSON.

Seiurus aurocapillus (Linn.).
Oven-bird.

ORDER PASSERES. PERCHING BIRDS. (*Continued.*)

SUBORDER OSCINES. SONG BIRDS. (*Continued.*)

FAMILY **MNIOTILTIDÆ.** WOOD WARBLERS. (*Continued.*)

GENUS SEIURUS SWAINSON. (*Continued.*)

Seiurus noveboracensis (Gmel.).
Water-Thrush.

Seiurus noveboracensis notabilis (Ridgw.).
Grinnell's Water-Thrush.

Seiurus motacilla (Vieill.).
Louisiana Water-Thrush.

GENUS GEOTHLYPIS CABANIS.

Geothlypis formosa (Wils.).
Kentucky Warbler.

Geothlypis agilis (Wils.).
Connecticut Warbler.

Geothlypis philadelphia (Wils.).
Mourning Warbler.

Geothlypis trichas (Linn.).
Maryland Yellow-throat.

Geothlypis trichas ignota Chapm.
Florida Yellow-throat.

GENUS ICTERIA VIEILLOT.

Icteria virens (Linn.).
Yellow-breasted Chat.

GENUS WILSONIA BONAPARTE.

Wilsonia mitrata (Gmel.).
Hooded Warbler.

Wilsonia pusilla (Wils.).
Wilson's Warbler.

Wilsonia canadensis (Linn.).
Canadian Warbler.

GENUS SETOPHAGA SWAINSON.

Setophaga ruticilla (Linn.).
American Redstart.

FAMILY **MOTACILLIDÆ.** WAGTAILS.

GENUS MOTACILLA LINNÆUS.

Motacilla alba Linn.
White Wagtail.

GENUS ANTHUS BECHSTEIN.

Anthus pensilvanicus (Lath.).
American Pipit.

Anthus pratensis (Linn.).
Meadow Pipit.

Anthus spragueii (Aud.).
Sprague's Pipit.

FAMILY **TROGLODYTIDÆ.** WRENS, THRASHERS, ETC.

SUBFAMILY **MIMINÆ.** THRASHERS.

GENUS MIMUS BOIE.

Mimus polyglottos (Linn.).
Mockingbird.

GENUS GALEOSCOPTES CABANIS.

Galeoscoptes carolinensis (Linn.).
Catbird.

GENUS HARPORHYNCHUS CABANIS.

Harporhynchus rufus (Linn.).
Brown Thrasher.

SUBFAMILY **TROGLODYTINÆ.** WRENS.

GENUS THRYOTHORUS VIEILLOT.

Thryothorus ludovicianus (Lath.).
Carolina Wren.

Thryothorus ludovicianus miamensis Ridgw.
Florida Wren.

Thryothorus bewickii (Aud.).
Bewick's Wren.

GENUS TROGLODYTES VIEILLOT.

Troglodytes aëdon Vieill.
House Wren.

Troglodytes aëdon aztecus Baird.
Western House Wren.

Troglodytes hiemalis Vieill.
Winter Wren.

GENUS CISTOTHORUS CABANIS.

Cistothorus stellaris (Licht.).
Short-billed Marsh Wren.

Cistothorus palustris (Wils.).
Long-billed Marsh Wren.

Cistothorus palustris griseus Brewst.
Worthington's Marsh Wren.

Cistothorus marianæ Scott.
Marian's Marsh Wren.

FAMILY **CERTHIIDÆ.** CREEPERS.

GENUS CERTHIA LINNÆUS.

Certhia familiaris americana (Bonap.).
Brown Creeper.

Bird Studies.

ORDER PASSERES. PERCHING BIRDS. (*Continued.*)

SUBORDER OSCINES. SONG BIRDS. (*Continued.*)

FAMILY **PARIDÆ.** NUTHATCHES AND TITS.

SUBFAMILY **SITTINÆ.** NUTHATCHES.

GENUS SITTA LINNÆUS.

Sitta carolinensis Lath.
White-breasted Nuthatch.

Sitta carolinensis atkinsi Scott.
Florida White-breasted Nuthatch.

Sitta canadensis Linn.
Red-breasted Nuthatch.

Sitta pusilla Lath.
Brown-headed Nuthatch.

SUBFAMILY **PARINÆ.** TITMICE.

GENUS PARUS LINNÆUS.

Parus bicolor Linn.
Tufted Titmouse.

Parus atricapillus Linn.
Chickadee.

Parus carolinensis Aud.
Carolina Chickadee.

Parus hudsonicus Forst.
Hudsonian Chickadee.

FAMILY **SYLVIIDÆ.** WARBLERS, KINGLETS, GNATCATCHERS.

SUBFAMILY **REGULINÆ.** KINGLETS.

GENUS REGULUS CUVIER.

Regulus satrapa Licht.
Golden-crowned Kinglet.

Regulus calendula (Linn.).
Ruby-crowned Kinglet.

SUBFAMILY **POLIOPTILINÆ.** GNATCATCHERS.

GENUS POLIOPTILA SCLATER.

Polioptila cærulea (Linn.).
Blue-gray Gnatcatcher.

FAMILY **TURDIDÆ.** THRUSHES, SOLITAIRES, STONECHATS, BLUEBIRDS, ETC.

SUBFAMILY **TURDINÆ.** THRUSHES.

GENUS TURDUS LINNÆUS.

Turdus mustelinus Gmel.
Wood Thrush.

Turdus fuscescens Steph.
Wilson's Thrush.

Turdus fuscescens salicicola (Ridgw.).
Willow Thrush.

Land Birds of Eastern North America.

ORDER PASSERES. PERCHING BIRDS. (*Continued.*)

SUBORDER OSCINES. SONG BIRDS. (*Continued.*)

FAMILY **TURDIDÆ.** THRUSHES, SOLITAIRES, STONECHATS, BLUEBIRDS, ETC. (*Continued*)
SUBFAMILY **TURDINÆ.** THRUSHES. (*Continued.*)
GENUS TURDUS LINNÆUS. (*Continued.*)

Turdus aliciæ Baird.
    Gray-cheeked Thrush.

Turdus aliciæ bicknelli (Ridgw.).
    Bicknell's Thrush.

Turdus ustulatus swainsonii (Cab.).
    Olive-backed Thrush.

Turdus aonalaschkæ pallasii (Cab.).
    Hermit Thrush.

Turdus iliacus Linn.
    Red-winged Thrush.

GENUS MERULA LEACH.

Merula migratoria (Linn.).
    American Robin.

GENUS HESPEROCICHLA BAIRD.

Hesperocichla nævia (Gmel.).
    Varied Thrush.

GENUS SAXICOLA BECHSTEIN.

Saxicola œnanthe (Linn.).
    Wheatear.

GENUS SIALIA SWAINSON.

Sialia sialis (Linn.).
    Bluebird.

INDEX TO COMMON AND SCIENTIFIC NAMES.

Figures in roman type refer to body of book.　　*Figures in italic type refer to the Systematic Table of Birds.*

BOOKS FOR THE COUNTRY.

Landscape Gardening. Notes and Suggestions on Lawns and Lawn-Planting, Laying Out and Arrangement of Country-Places, Large and Small Parks, Cemetery Plots, and Railway-Station Lawns; Deciduous and Evergreen Trees and Shrubs, The Hardy Border, Bedding Plants, Rockwork, etc. By SAMUEL PARSONS, Jr., Superintendent of Parks, New York City. With nearly 200 illustrations. Large 8°, $3.50.

" Mr. Parsons proves himself a master of his art as a landscape gardener, and this superb book should be studied by all who are concerned in the making of parks in other cities."
—*Philadelphia Bulletin.*

Lawns and Gardens. How to Beautify the Home Lot, the Pleasure Ground, and Garden. By N. JÖNSSON-ROSE, of the Department of Public Parks, New York City. With 172 plans and illustrations. Large 8°, gilt top, $3.50.

This book is intended to be a help to all lovers of Gardening. It treats of the practical side of landscape gardening, describes the best hardy plants, and points out the proper use of each.

Wild Flowers of the Northeastern States. Drawn and carefully described from life, without undue use of scientific nomenclature, by ELLEN MILLER and MARGARET C. WHITING. With 308 illustrations the size of life, and Frontispiece. Large quarto, buckram, 8½ x 12¼ inches. New edition in smaller form, 8°,
net, $3.00

This work presents upwards of 308 drawings of American wild flowers, and careful descriptions of the flowers so depicted, and covers ground which has not been covered by any previous botanical publication. In every case great care has been taken to depict the peculiar traits, the average size, and all the details of each individual plant.

Among the Moths and Butterflies. By JULIA P. BALLARD, author of "Building Stories," etc. 8°, pp. xxxiv. + 237, $1.50.

" The book, which is handsomely illustrated, is designed for young readers, relating some of the most curious facts of natural history in a singularly pleasant and instructive manner."— *N. Y. Tribune.*

The Trees of Northeastern America. By CHARLES S. NEWHALL. With an Introductory Note by NATH. L. BRITTON, E.M., Ph.D., of Columbia College. With illustrations made from tracings of the leaves of the various trees. New issue, 8°, reduced to $1.75.

" We believe this is the most complete and handsome volume of its kind, and on account of its completeness and the readiness with which it imparts information that everybody needs and few possess, it is invaluable."— *Binghamton Republican.*

The Shrubs of Northeastern America. By CHARLES S. NEWHALL, author of " The Trees of Northeastern America," etc. Fully illustrated. New issue, 8°, reduced to $1.75.

" This volume is beautifully printed on beautiful paper, and has a list of 116 illustrations calculated to explain the text. It has a mine of precious information, such as is seldom gathered within the covers of such a volume."
—*Baltimore Farmer.*

The Vines of Northeastern America. By CHARLES S. NEWHALL. Fully illustrated. New issue, 8°, reduced to $1.75.

" The work is that of the true scientist, artistically presented in a popular form to an appreciative class of readers."—*The Churchman.*

The Leaf Collector's Handbook and Herbarium. An aid in the preservation and in the classification of specimen leaves of the trees of Northeastern America. By CHARLES S. NEWHALL. Illustrated. 8°, pp. xv. + 203, $2.00.

" The idea of the book is so good and so simple as to recommend itself at a glance to everybody who cares to know our trees or to make for any purpose a collection of their leaves."—*N. Y. Critic.*

The Wonders of Plant Life. By Mrs. S. B. HERRICK. 16°, beautifully illustrated. $1.50.

The only thing aimed at is to give the more important types in a popular way, avoiding technicalities where ordinary language could be substituted, and, where it could not, giving clear explanations of the terms.

" A dainty volume . . . opens up a whole world of fascination . . . full of information."—*Boston Advertiser.*

G. P. PUTNAM'S SONS, 27 & 29 West 23d Street, N. Y.

Books on Hunting.

www.ingramcontent.com/pod-product-compliance
Lightning Source LLC
Chambersburg PA
CBHW021357210326
41599CB00011B/919